1001

WONDERS OF THE UNIVERSE

1001
WONDERS OF THE UNIVERSE

PIERS BIZONY

Quercus

CONTENTS

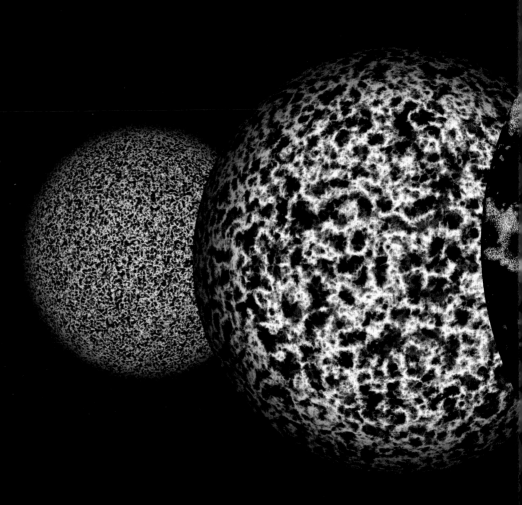

THE EXPANDING UNIVERSE
Present theory dates the origin of the universe at around 13.7 billion years ago,
the time of the Big Bang. Since then it has been expanding and cooling. The stages
illustrated show the universe as it may have been at 10 billion (left), 5 billion, one
billion and 200 million years ago.

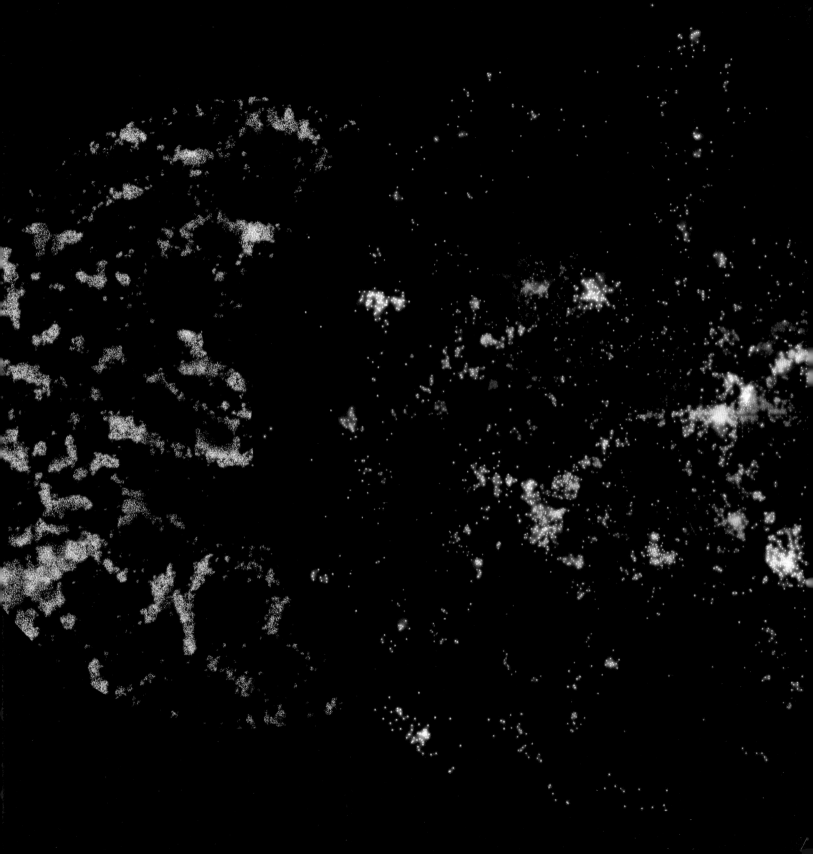

INTRODUCTION

Humankind's relationship with the stars began long before the first astronomical records were made on carved stones and tablets, in papyrus scrolls or leather – bound volumes.

More than 4000 years ago, the Mesopotamian civilizations, living in what is now called Iraq, knew how to navigate by the stars and predict the movements of five prominent sparks, 'wanderers', whose movements were not the same as the other stars. In the absence of telescopes, it is impressive how many ancient thinkers speculated that the bright sparkles in the night sky might be other suns, and even that countless worlds similar to ours could exist, with air and sea, and inhabitable lands. Even so, very few thinkers doubted Earth's importance as the very heart of the cosmos.

Around 1,870 years ago, a Greek mathematician called Ptolemy, in the Egyptian city of Alexandria, reinforced the popular assumption that the sun, stars and planets orbited the Earth. He devised a useful geometric description that held sway for the next 1400 years. Stars and planets were arranged in nested spheres rotating around the Earth. Islamic scholars took up his work, and eventually, as Christian dogma took hold across Medieval and early Renaissance Europe, it became not just absurd but dangerous to suggest that the human realm could be anything else but at the heart of all existence.

By the 16th century, disciplined students of the stars began slowly to disentangle their science from the ancient confusions of astrology. There was widespread respect for Ptolemy, but as new measurements and better timings of events in the night sky became available, too many discrepancies emerged between his tidy schemes and the more complicated data revealed by actual observations. A better theory was needed to explain all the movements of the sun, stars and planets across the sky.

And still, after all those thousands of years, no one had ever seen any of these wonders except as twinkling smudges of light. Even the moon's terrain was a mystery. In 1610, Galileo Galilei directed one of the earliest telescopes towards Jupiter. He saw its moons, which was stunning in itself. Of even greater significance was the way that sunlight illuminates the sphere of Venus, sometimes full-on and at other times, obliquely, just as is the case with our own moon.

From these observations, Galileo claimed that the sun could not be orbiting the Earth. Rather, the Earth, and all other known planets, had to be orbiting the sun. This could only mean that we are not the centre of the universe: an idea quite literally unthinkable to the Church. Galileo spent the rest of his life in danger of torture and execution for heresy – but his writings escaped into the wider world, and our understanding of our place in the cosmos was changed forever.

Despite rapid improvements in telescope technology, yet another three centuries passed before we could see anything much more than the mountains of the moon, the blurry discs of the most prominent planets or the smudgy tails of comets. How privileged we are, in just these most recent decades of the two million-year long human journey, to witness a time of space probes that can actually visit some of these worlds and scan them from close by – or even land upon them. How amazing that our orbiting space telescopes can operate far beyond the hazy interference of the Earth's atmosphere. Quite literally we can see more of the cosmos than ever before.

In a thousand and one images, each just as fascinating as any of Sheherezade's thousand and one arabian tales, this book takes us on a journey through the solar system, into the vast and terrifying emptiness between stars, and to the other side of our galaxy, the Milky Way. Then we travel not just through space but back in time, not just to the dawn of the cosmos, but its first few fractions of a second.

As this book encounters the outermost limits of our present-day vision, so we reach the equivalent boundaries of current scientific understanding. The more we find out, the more puzzles confront us. Stars and galaxies alone cannot account for all the gravity that holds the cosmos together. Unfamiliar facets of matter and energy are at work on the grandest scales. The Big Bang that gave rise, in a single instant some 13.75 billion years ago, to all of existence has yet to be fully understood, and it may even be that 'our' observable universe is just one among many in an even grander 'multiverse'. Future generations will be just as enthralled by their astronomical discoveries as we are in these times. Cosmic phenomena that might seem impossible to us today will shape the new perspectives of tomorrow.

Piers Bizony

SYSTEM

All but three percent of our Solar System's mass is concentrated in the Sun. Eight planets, Mercury, Venus, Earth, Mars, Jupiter, Saturn, Uranus and Neptune, travelling in near-circular orbits, account for most of the rest.

The planets are confined within a flat disc-shaped region of space, the Plane of the Ecliptic. All planets apart from Mercury and Venus have other, much smaller worlds, called moons, orbiting around them. Moons range in shape and scale from about the size and roundness of our Earth's familiar moon, to jagged chunks of rock just a few kilometres wide. More than 140 moons have been discovered so far. A ninth 'planetary' world, Pluto, is so small it does not count as a proper planet. It is so far away from the Sun that it takes 248 Earth years to complete an orbit. Far beyond Pluto, scattered islands of dust, gas and cometary chunks of ice roam the Solar System's dark perimeter. Light from the dwindling Sun travels for at least a year or two before it glints, faintly, on these last, lonely outposts. Beyond that point, the Sun's influences cannot be told apart from those of the two hundred billion other stars whose collective gravity binds the Milky Way galaxy together.

Around five billion years ago, a vast cloud of dust and gas drifted through our region of the galaxy. These thin, widely scattered wisps originated from supernovae among an earlier generation of stars. The cloud, called a nebula, was gently pulled around by gravitational forces from other, surviving stars in the surrounding galaxy. Most of the atoms in the nebula were hydrogen or helium, but all the other chemical elements were present in smaller amounts. Somewhere within the nebula, a seemingly insignificant clump of matter coalesced, gaining mass until it was capable of exerting its own gravitational force, extremely weak, yet just strong enough to attract nearby atoms and molecules. This mass, known by scientists as a protostar, grew to the point where its gravity became strong enough to pull in more material from ever greater distances: tens, then thousands, and eventually many millions of kilometres. Dust and gas from the surrounding nebula became concentrated into a spinning plate-shaped structure known as an accretion disc, with the protostar, our young Sun, at its centre.

Even as the Sun gathered mass, some of the debris in the accretion disc began to cluster in miniature, and much cooler, versions of the protostar process. Small worlds, known as planetesimals, began to form. When these collided with each other and their wreckage recombined, these larger structures became the planets and moons that exist today. One of those worlds became the Earth. It began life around 4.6 billion years ago as a hot, chaotic sphere of molten metals and silicates, but over time the various materials began to divide according to density, with the heaviest metals, such as iron, sinking into the Earth's core, and the lighter materials, such as silica, floating upwards towards the planet's surface. After approximately 100 million years, the outer surface cooled sufficiently for a thin rocky crust to form.

The Sun has burned for five billion years, and will burn for five billion more, after which it will have used up a crucial proportion of its available hydrogen. The local star that has nurtured us for so long will turn into a violent and unpredictable monster prone to sudden nuclear shutdowns and reignitions, expansions and contractions, alternating over several millennia. It will end its days as a red giant, so gigantic that its wayward mantle of gases will swallow up half the planets of the Solar System, including the Earth. Mercury and Venus will simply vanish. Everything on our planet that lives and breathes will be utterly vaporized. The grass, the trees, even the toughened lichen that clings to granite will be blasted into atoms. The oceans will boil away to the last drop. The ice caps will vaporize. The sands on every beach in the world will fuse into glass. If Earth survives at all, it will do so only as a scorched lump of rock. Fortunately this disaster lies so far in the future there is no point worrying about it.

THE CANYONS OF MARS *The largest canyon in the Solar System cuts a wide swath across the face of Mars. Named Valles Marineris, the system is more than 3,000 km (1,860 miles) long and up to 8 km (5 miles) deep. By comparison, the Earth's Grand Canyon in Arizona is the merest scratch. The origins of this dramatic scar are uncertain. The Martian crust may have cracked billions of years ago as the planet cooled.*

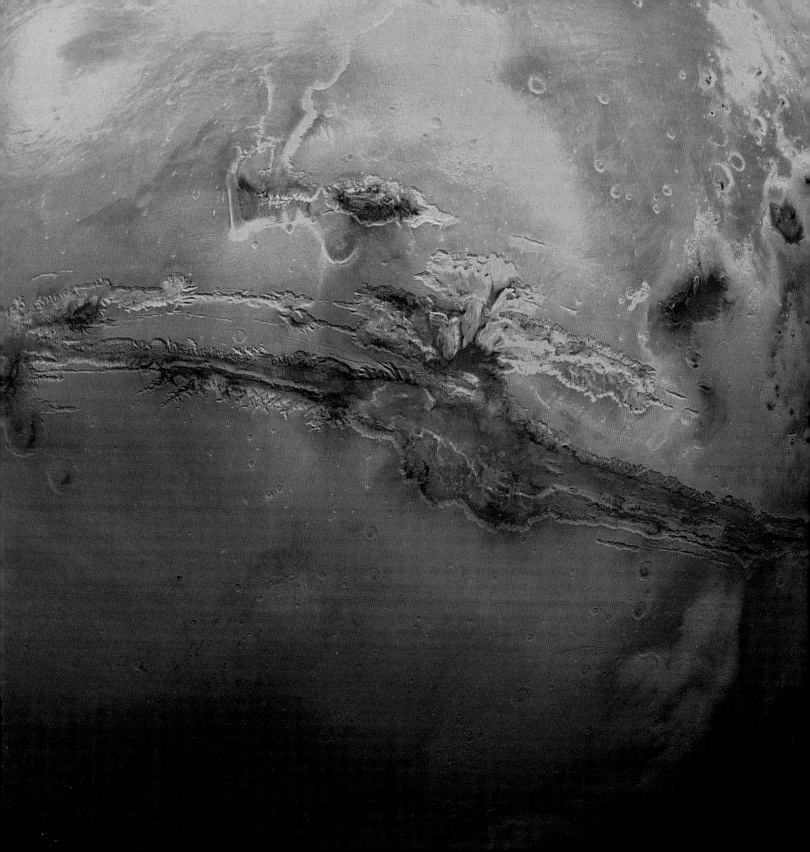

OUR STAR, THE SUN

In the 4.5 billion years since its birth, the Sun has converted four million tons of hydrogen into energy every second, yet the amount of available hydrogen is still so great that it should be able to sustain this process for another 5,000 million years to come.

In the Sun's core, immense gravitational pressure squeezes hydrogen nuclei together until helium is formed, in a process known as fusion. The mass of helium that emerges from this process is 0.7 per cent less than the mass of hydrogen that goes into it. That 'missing' mass is liberated in the form of heat and infra-red energy, ultraviolet and visible light, and a small proportion of X rays and radio waves. What we call 'sunshine' takes a mere eight minutes to reach as far as the Earth.

Temperatures in the Sun's atmosphere, known as the Corona, are at least two million degrees centigrade, far greater than those at the Sun's surface (the photosphere). The corona is threaded with powerful and fast-changing magnetic fields that constantly twist and distort into new patterns. Positively and negatively-charged subatomic particles spiral at very high energies along the magnetic field lines.

Vast clouds of electrically charged subatomic particles are continually blasted away from the Sun. These violent bursts, which often eject the equivalent of 20 billion tons of matter, are properly known as Coronal Mass Ejections (CMEs). The swarm of particles travel across interplanetary space, interacting with all the planets and Moons in the solar system, including the Earth. The powerful electrical charges in the 'solar wind' can disrupt computers, interfere with national power supplies and totally destroy the circuitry of orbiting satellites.

Sunspots, the familiar dark-looking blemishes on the Sun's otherwise bright disc, are concentrated regions near the surface where magnetic fields become extremely strong. Sunspots usually last for several days, although large ones may survive for several weeks. Sunspots can be seen when the eyepiece end of a telescope is focused onto a sheet of white paper, but only a fool would actually look into the telescope. All through recorded history, we humans have known the costs of staring directly at the Sun.

THE SUN *On a clear afternoon the Sun is a dazzlingly bright disc of featureless white light: so bright we cannot safely look at it with the naked eye. A closer look through strong filters shows the Sun's spherical shape and its complex and ever shifting structures, such as solar flares and sunspots.*

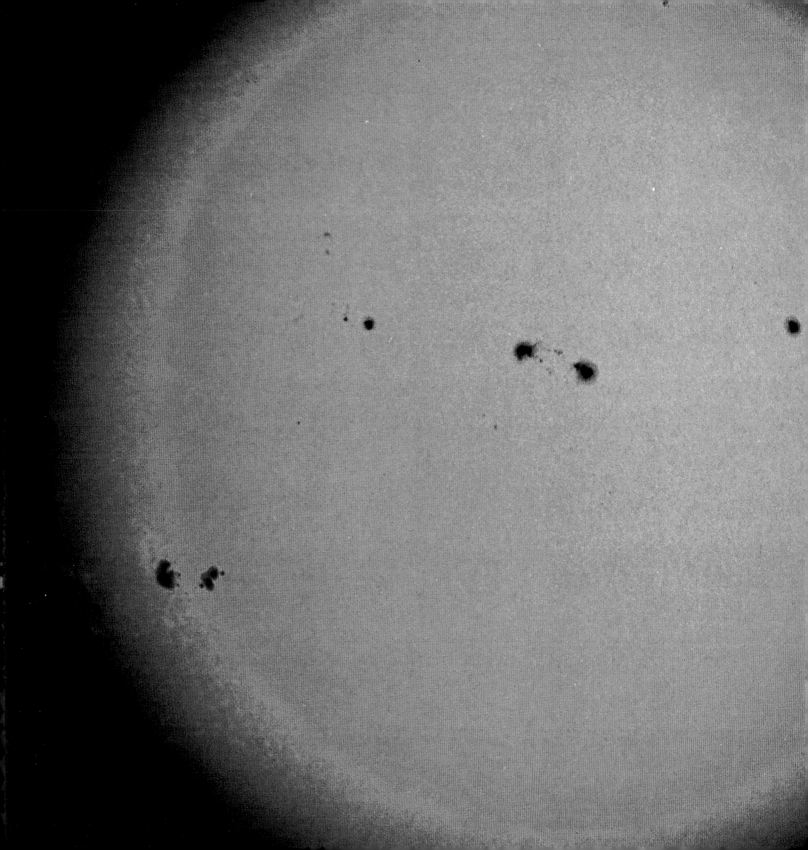

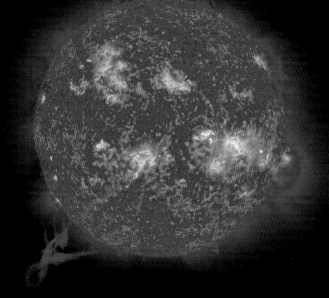
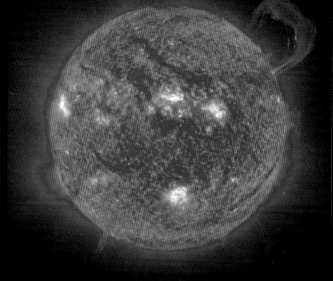

1 Solar storms

Coronal Mass Ejections, or CMEs, can hurl more than 100 billion tons of mass into deep space at speeds of millions of kilometres per hour. The ejected mass is in the form of subatomic particles. When these reach the Earth, they interact with the planet's upper atmosphere and magnetic field.

2, 3 Solar prominences

Vast curling clouds of relatively cool plasma can easily be seen through a telescope with a suitably dense filter to shut out the glare. Unlike CMEs, these visually dramatic prominences are usually confined within the Sun's atmosphere, the corona.

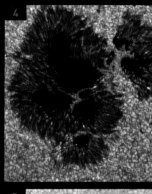

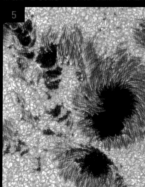

4, 5 Sunspots

The photosphere is the obviously bright surface of the Sun. It is often dappled with sunspots, created when huge and powerful loops within the Sun's magnetic field force their way through the photosphere.

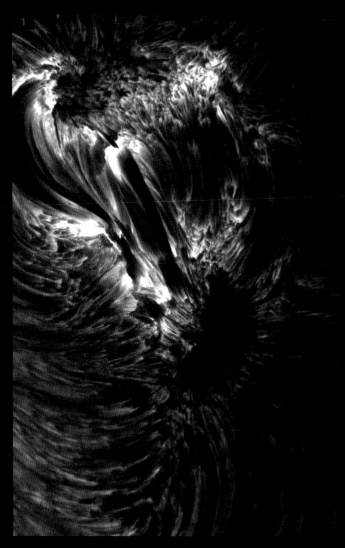

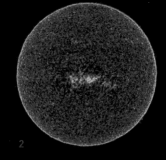

2 Solar grains

An image taken in extreme ultraviolet light reveals a seething mass of dark cells, a phenomenon known as granulation. Each grain is about the size of Texas. Granulation is caused by convection currents inside the Sun. Hot material rises into the photosphere, then cools down (if anything on the Sun can be called cool) and falls back around the edges of each cell.

3 The magnetic Sun

Thus image, taken by the Solar Heliospheric Observatory (SOHO) satellite, shows gas at different temperatures, and various frequencies of radiation emanating from the Sun, along with a tangle of twisted streamers and prominences reaching into the corona and beyond. We are a long way from fully understanding how the Sun's magnetic field shapes its complicated behaviour.

1 The heart of a sunspot

This extreme close-up is one of the sharpest images ever obtained of a sunspot. It reveals features as small as 100 km (60 miles) across. Dark thread-like cores of plasma run the length of the bright filaments, all shaped by the restless rearrangements of the magnetic field.

4 Space station

An Earth telescope captures the International Space Station silhouetted as it passes in front of the Sun's disc. The Station's shadow is not to scale with the Sun, because the Station orbits very close to the Earth. The intersection lasted just fractions of a second. Careful timing was needed to capture this shot.

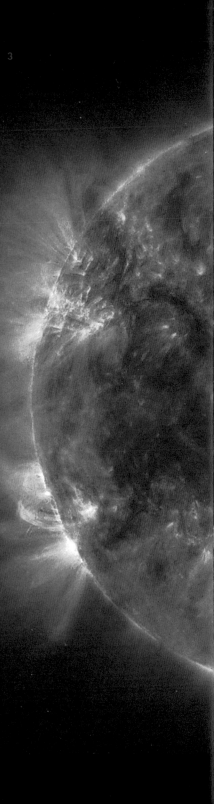

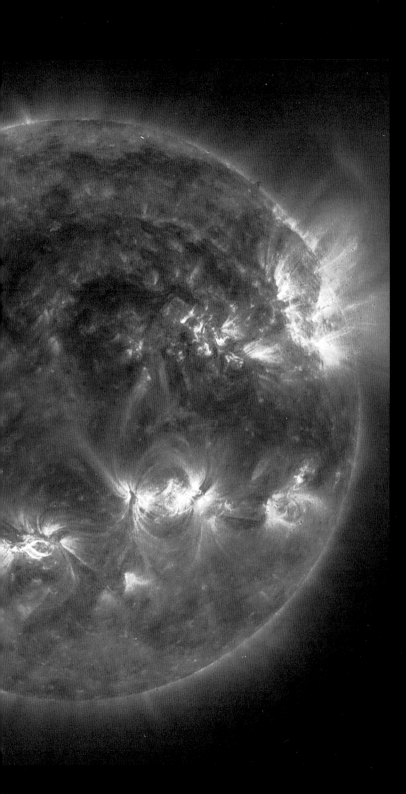

5 ECLIPSE

A solar eclipse occurs when the Moon interposes itself between the Sun and the Earth, temporarily obscuring our view of the Sun. By a lucky coincidence, the Moon's disc neatly superimposes over the Sun's disc during full eclipse, leaving just the corona visible.

6 ENERGETIC THREADS

Suspended by magnetic fields above an active region of the photosphere, these dark filaments span more than 40 times the diameter of the Earth. Such filaments are unstable and impermanent. The scene was captured in extreme ultraviolet by the Solar Dynamics Observatory, one of many space-based systems monitoring the Sun.

7 JETS ON THE SUN

Imagine a pipe as wide as Arizona and as tall as the Earth from pole to pole. Now imagine this pipe filled with hot gas moving at 50,000 km (30,000 miles) per hour. Think of the pipe made not from any solid materials but just by a transparent magnetic field. This is a spicule, one of thousands constantly generated by the Sun. Spicules last for around five minutes, then fade.

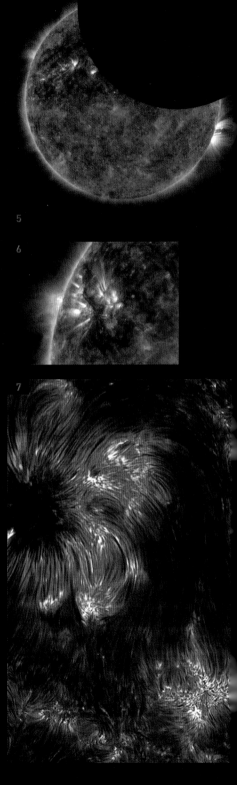

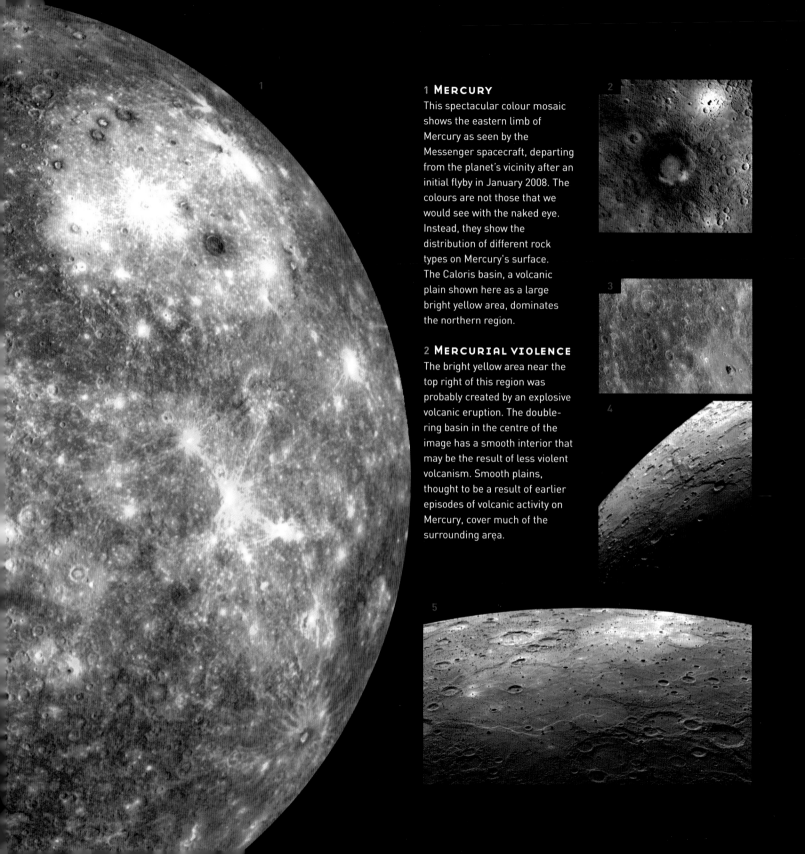

1 MERCURY
This spectacular colour mosaic shows the eastern limb of Mercury as seen by the Messenger spacecraft, departing from the planet's vicinity after an initial flyby in January 2008. The colours are not those that we would see with the naked eye. Instead, they show the distribution of different rock types on Mercury's surface. The Caloris basin, a volcanic plain shown here as a large bright yellow area, dominates the northern region.

2 MERCURIAL VIOLENCE
The bright yellow area near the top right of this region was probably created by an explosive volcanic eruption. The double-ring basin in the centre of the image has a smooth interior that may be the result of less violent volcanism. Smooth plains, thought to be a result of earlier episodes of volcanic activity on Mercury, cover much of the surrounding area.

3, 4, 5 MERCURY

A large crater has unusual dark material [3] of unknown composition near its centre. One possibility is that the material was uncovered during the impact that created the surrounding crater. Certainly Mercury has been struck on many occasions by asteroids, which have left many craters [4]. The planet's typical landscape [5] is a mix of impact craters and terrain shaped by volcanism.

6 HOSTILE MERCURY

Tiny, fast-moving Mercury is a moon-like world seared by the heat of the Sun, and ravaged by impacts from asteroids drawn towards it by the Sun's remorseless gravitational strength. The sunlit surface is hot enough to melt lead.

6

7

7 MYSTERIOUS TERRAIN

Less than half the planet's surface has been mapped in detail. This composite image shows what looks like an unusual streak of featureless terrain stretching from south to north. In fact, this is a visual representation of missing data.

AN 'EARTH' GONE WRONG?

Of all the planets in the Solar System apart from the Earth, one in particular carries a stark warning about what happens when environmental dynamics run out of control.

Inappropriately named for the ancient goddess of love, Venus is an unfriendly world permanently clouded by a dense carbon dioxide atmosphere. None of its surface features are visible through conventional telescopes. Radar scans reveal energetic volcanism and mountains deformed by heat. The atmosphere traps heat from the Sun so that a runaway 'greenhouse' effect has taken hold, making this a hellish place where ground temperatures are high enough to melt lead.

In December 1970, Soviet Russia achieved a successful landing of a robotic spacecraft on Venus. The lander survived for about twenty minutes before succumbing to the intense heat and pressure, but it lasted just long enough to send the first pictures ever taken on the surface of another planet.

HELLISH VENUS *In visible light, Venus appears as a featureless sphere, similar in size to the Earth. The planet's surface is entirely obscured by a thick atmosphere of carbon dioxide. This seemingly calm gaseous shell veils a world of hellish temperatures and nightmarish landscapes swathed in clouds of sulphuric acid.*

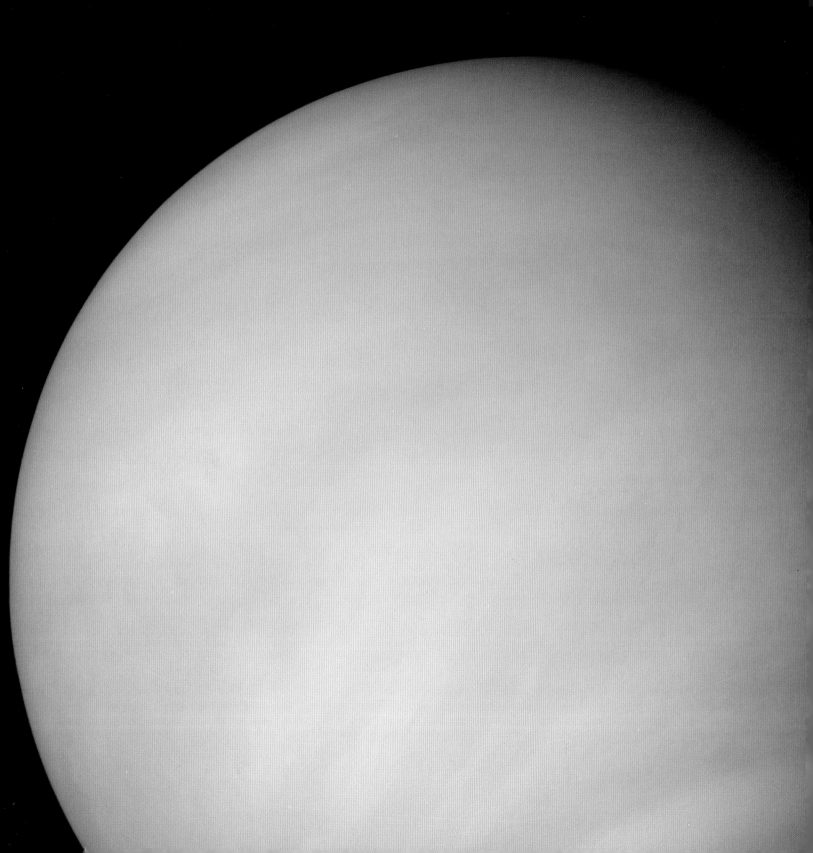

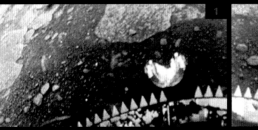

3 VENUS VIA RADAR

In the early 1990s, NASA's Magellan space probe penetrated the atmosphere of Venus with radar instruments, gradually building up a height map of the surface terrain, including large mountains.

1, 2 TOUCHDOWN ON VENUS

The first pictures of the Venusian surface were transmitted to Earth by Russia's Venera 13 robotic lander in 1982. Flat rocks are scattered on a dark soil amidst a volcanic landscape. The probe survived for just two hours before the hostile surface conditions disabled it.

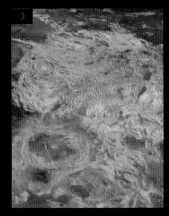

4 VENUSIAN VOLCANO

This radar image shows the Maat Mons volcano. Its peak is covered by rough lava flows that wind like rivers across the surrounding lowlands of older and smoother terrain.

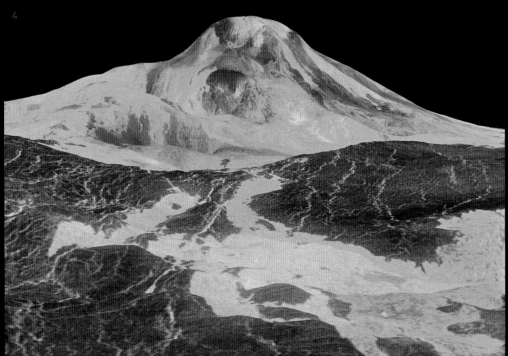

5 VENUSIAN ENVIRONMENT

Venus is named after the Roman goddess responsible for matters of love, but nothing could be less lovely than this world, with its surface shaped by volcanoes, lava flows and the corrosive effects of its noxious acid-misted atmosphere.

1, 2, 3 Global Venus

These hemispheric visualisations were compiled from ten years' worth of data from different space probes and radar instruments. In a view centred at 270 degrees east longitude [1] the purple areas represent the lowest-lying regions of the crust, as measured from the centre of the planet. The bright white area towards the bottom of this scan from directly above the planet's North Pole [2] is Maxwell Montes, the highest mountain on Venus. A mosaic centred at 180 degrees east longitude [3] shows the volcanic nature of the forces shaping the crust.

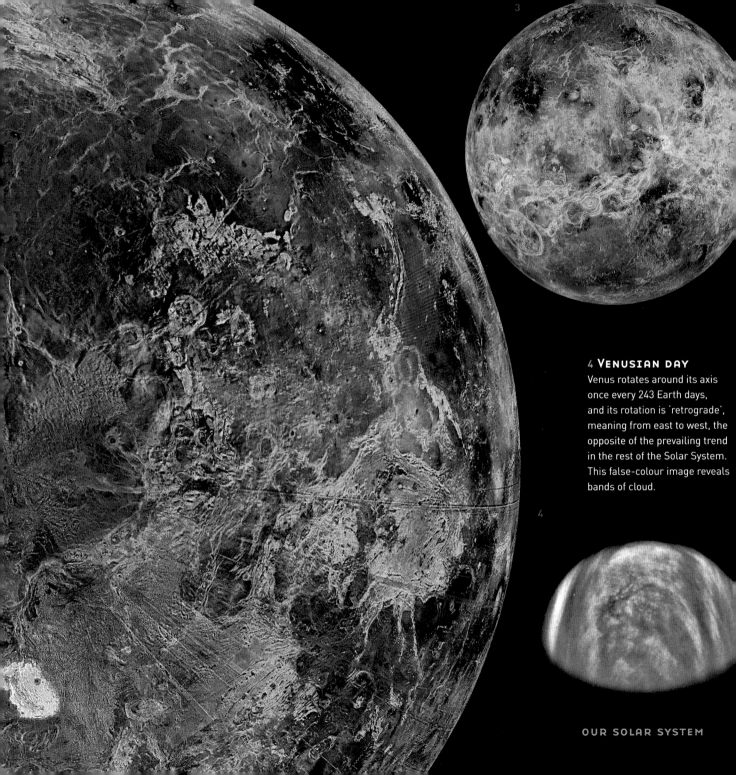

4 VENUSIAN DAY
Venus rotates around its axis
once every 243 Earth days,
and its rotation is 'retrograde',
meaning from east to west, the
opposite of the prevailing trend
in the rest of the Solar System.
This false-colour image reveals
bands of cloud.

1, 2, 3 VENUSIAN DRAMA

Radar mapping makes these pancake-like features [1] look like craters, but they are actually raised blisters, approximately 65 km (40 miles) across. They formed when thick lava flowed through vents under the crust, and set in place before the deformations could settle back to their original heights. Other features, such as this appropriately-named 'Arachnoid' [2] are circular fractures created as underlying lava forced the crust upwards, cracking it into patterns like spider webs. Coronae, sunken areas surrounded by roughly concentric rings [3], are formed in a similar process of crustal swelling and collapse.

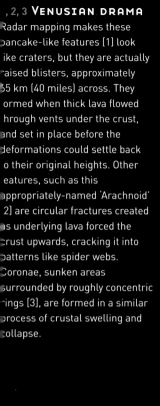

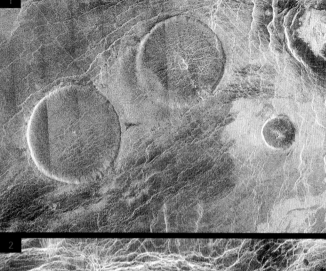

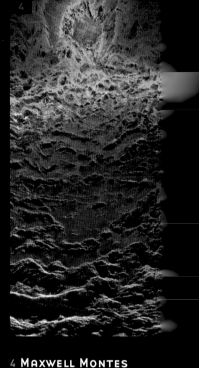

4 MAXWELL MONTES

This view westwards toward the slopes of Maxwell Montes, shows details of the tallest mountain on Venus. The dent-like feature on the slope of Maxwell is the impact crater Cleopatra.

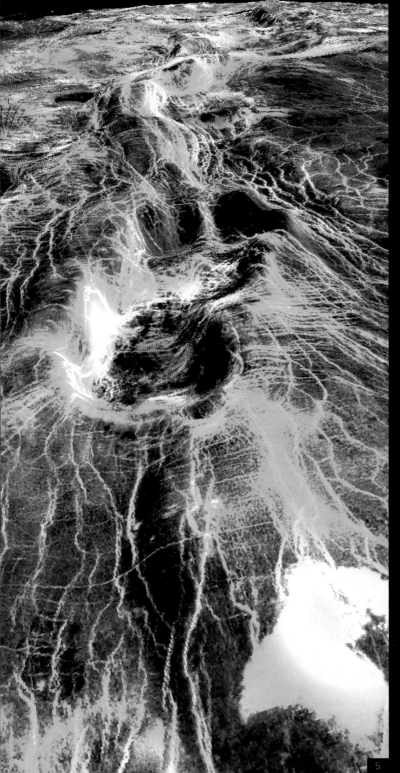

5

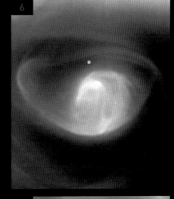

6

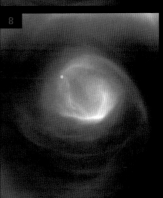

8

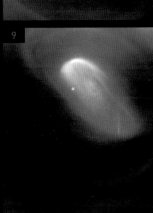

7

9

5 VENUSIAN PLAINS

This perspective computer-generated view, based on radar measurements, shows part of the lowland plains in Sedna Planitia. The circular depressions are coronae, which are apparently unique to the lowlands of Venus. Coronae tend to cluster in rows along the edges of the planet's major tectonic belts, the large sections of crust that 'float' on top of the underlying magma, somewhat like the Earth's great continental plates.

6, 7, 8, 9 WEATHER ON VENUS

Vortices of gas whirl around the polar regions of Venus, shaped by the planet's rotational energy. High-altitude gas from the equatorial region travels towards the poles, where it loses some of its heat and sinks back into the lower levels of the atmosphere. Similar, if much milder, processes shape the weather on Earth.

OUR HOME
WORLD

It would be more appropriate to call our home world 'Oceania' rather than 'Earth'. Less than a third of its surface is dry land. A lucky set of circumstances makes this planet hospitable to life.

Earth occupies an orbit known as the Goldilocks Zone, which is neither so close to the Sun that temperatures reach too high for liquid water to exist, nor so distant that they plunge so low that water freezes. Within this narrow band of special conditions, biology can function. A dense atmosphere protects the Earth's surface against excessive ultraviolet radiation from the Sun, while an abundance of water creates a vast realm in which the chemistry of life can be sustained. Are these circumstances unique, or do countless similar worlds exist in the universe?

The Moon, our nearest companion in space, was formed when the Earth was just 100 million years old. A wayward young planetary body, approximately the size of Mars, smashed into the Earth. The impact sent vast amounts of rubble into space, creating a ring-shaped swarm of debris around the Earth. As this coalesced, the Moon took shape. Its materials are a mix of contributions from the original stray collider and the Earth.

EARTH *The familiar continents are clearly visible in this beautiful natural-colour composite of planet Earth, derived from different sets of space satellite data. The Indian subcontinent is at the centre of this view, with the Persian Gulf and the Red Sea at middle left, and Africa at the left edge.*

1, 2, 3 CITIES AT NIGHT
Human technologies are evident
on planet-wide scales, especially
at night, when the lights of our
cities are aglow. Parts of Europe
and Africa are illuminated in the
top image [1], as are the bright
lights of Cairo and Alexandria,
Egypt on the Mediterranean
coast [2] or the US cities of
Mobile, New Orleans and
Houston [3] as the viewpoint
moves southeastward.
These shots were taken by
crew members aboard the
International Space Station.

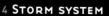

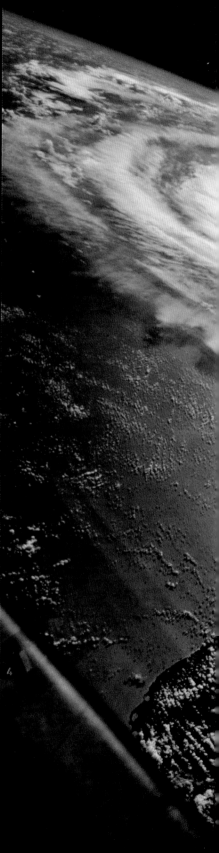

4 STORM SYSTEM
Hurricane Marilyn drifts
dangerously over Puerto Rico in
September 1995. The southern
half of Puerto Rico can be seen
outside the cloud cover, and
the island of Hispaniola is at
lower left.

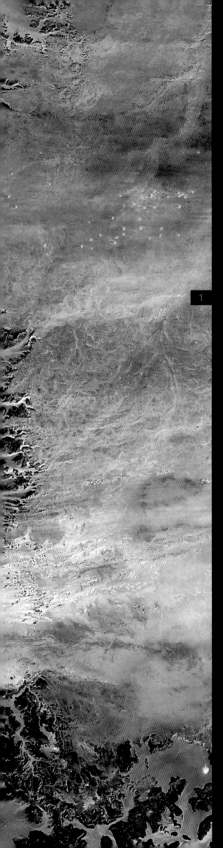

1 Dusty Earth
Deserts are a prominent part of the terrestrial landscape, covering more than a fifth of the dry surface area. This is part of the African Sahara, seen from orbit. Rocky granite outcrops push through the drifting sands.

2 Thin layer for life
The Earth's atmosphere seen edge-on in this image taken from the Space Station. Cloud formations are seen silhouetted against the setting Sun. The atmosphere is our only protection against the vacuum of space, and it does not amount to much. If the Earth were the size of an apple, the breathable layers of air would be no thicker than an apple's skin.

3 Land and sea
The west coast of Somalia in North Africa dominates this view, with the Gulf of Aden and the coast of Yemen at top, and the Indian Ocean at bottom right. No other planet in the solar system has liquid water flowing freely on its surface.

1 Sunlight meets the sky

Electrically charged subatomic particles from the Sun interact with the Earth's magnetic field and upper atmosphere, creating ghostly displays of light. On a clear night over the far northern areas of the world, it's easy to witness the eerie glow of an 'aurora' undulating in the sky. Similar displays are also visible in far southern polar regions.

2 Hidden stresses

Radar images taken from space at different times, and then overlapped as a single image, show stresses in the Earth's crust. The colour-coding reveals areas of crustal movement that have occurred between one scan date and the next. These tensions eventually resolve themselves in the form of earthquakes.

3 Extending the human realm

We are the first generation who can claim a permanent human presence away from the surface of the Earth. The International Space Station is always occupied by at least six people. One day we will establish colonies on the Moon and Mars.

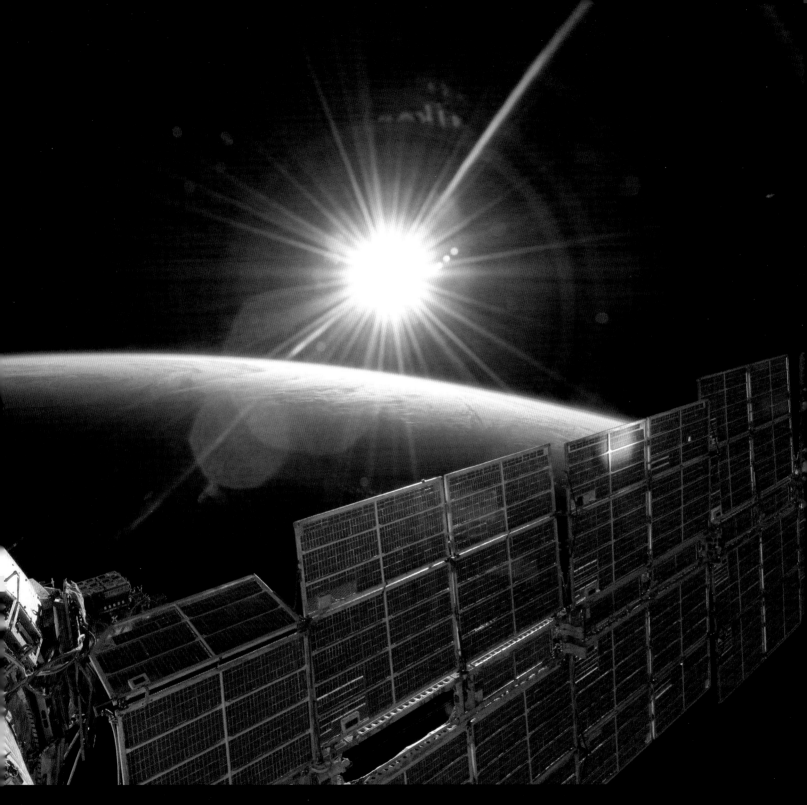

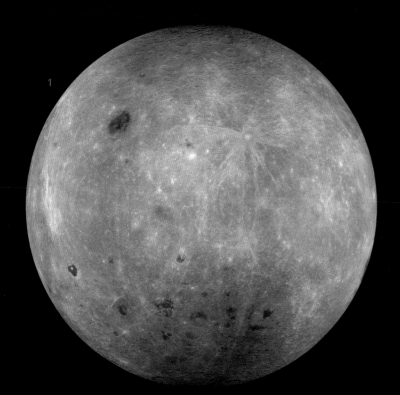

3 Familiar face

The near side of the Moon is the face that we always see. It is a mixture of smooth, dark terrain, called 'seas', and bright highlands. To the upper right of centre, the Seas of Tranquility and Serenity are visible, with the smaller oval of the Sea of Storms between them and the edge of the Moon's disc. The prominent impact crater towards the south is Tycho. The spokes shooting out in all directions consist of debris thrown out by the original impact.

1 Hidden face

The lunar far side is always turned away from the Earth. It is markedly dominated by highland terrain and crater impacts, with very few of the darker lowland 'seas' exhibited by the near side. Many people call this hidden face the 'dark side', but it receives just as much sunlight overall as the rest of the Moon does.

2 Moonrise from orbit

Space travellers get an excellent view of the Moon rising above the Earth's atmosphere, in this photo taken by the crew of a space shuttle orbiting the Earth once every ninety minutes. Through their windows, the Moon would seem to rise and set sixteen times a day.

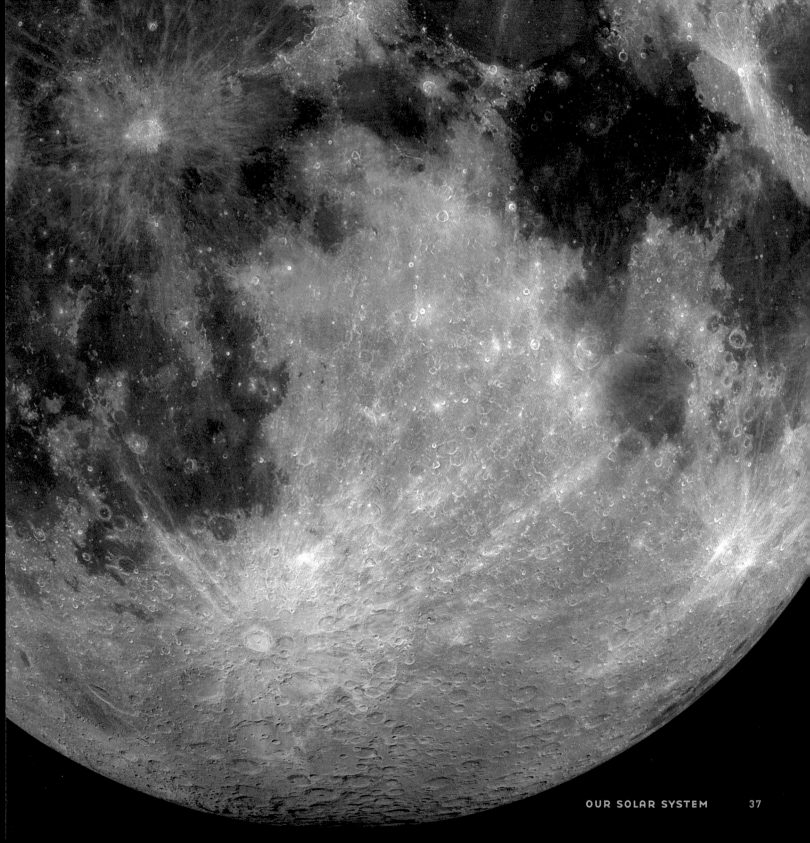

1 VIOLENT HISTORY

Molten lava created the flat 'seas' on the Moon's near side. The Earth's gravity shaped the lava, preventing much of it from flowing into the far side terrain. As a consequence, many ancient impact craters are still evident on the far side. The Earth was similarly pounded early in its history, but wind and weather have eroded most of the scars.

4 ETERNAL BOOTPRINT

Mankind's first steps on the Moon are preserved in the dust at the Apollo 11 landing site on the Sea of Tranquility. Future lunar visitors will need to respect the site as a historic landmark.

5 MOON MEN

Twelve humans have walked on the Moon's surface during the Apollo missions between 1969 and 1972. The prints of their boots in the lunar dust, and the tracks left by their wheeled rovers and carts, will survive for thousands of years, and perhaps even for tens of thousands.

2, 3 SECRET GLOW

The Moon's natural colour [2] is greyish white, but space-based instruments that reach beyond the scope of the human eye see other 'colours'. Image [3] shows a glow of x-rays generated by oxygen, magnesium, aluminium and silicon atoms in the lunar dust. The rays are caused by fluorescence when solar radiation bombards the surface.

6 APOLLO PANORAMA

This wide view, assembled from multiple overlapping photographs taken during the Apollo 17 mission in December 1972, shows astronaut Harrison Schmitt and his battery-powered lunar roving vehicle at the edge of a small, deep crater, nicknamed Shorty. This was the last manned lunar mission, but we may return there in years to come.

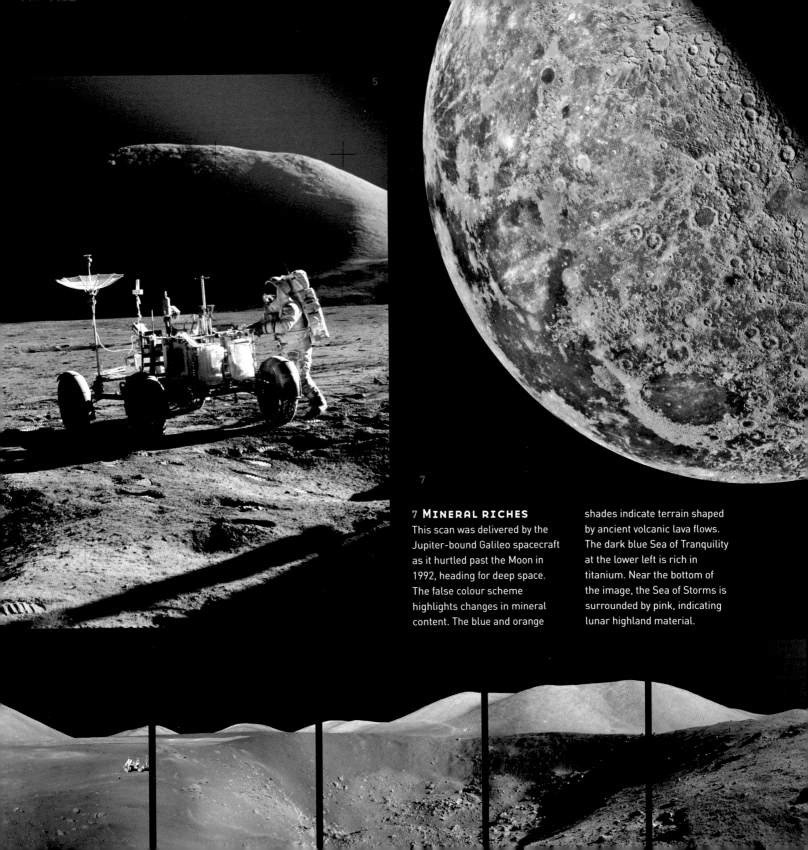

7 Mineral riches

This scan was delivered by the Jupiter-bound Galileo spacecraft as it hurtled past the Moon in 1992, heading for deep space. The false colour scheme highlights changes in mineral content. The blue and orange shades indicate terrain shaped by ancient volcanic lava flows. The dark blue Sea of Tranquility at the lower left is rich in titanium. Near the bottom of the image, the Sea of Storms is surrounded by pink, indicating lunar highland material.

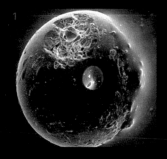

1 LUNAR BEAD

When a meteorite strikes the Moon, the energy of the smash melts some of the splattering rock, a fraction of which cools into tiny glass beads, known as spherules. Many of these were found in lunar soil samples returned to Earth by the Apollo astronauts. This spherule, no larger than a speck of sand, is particularly interesting because it has been the victim of an even smaller impact. A miniature crater is visible on the upper left, surrounded by a fragmented area caused by the shockwaves of the small impact.

2 MOUNTAINS AND VALLEYS

Hadley Rille looks like a meandering river, but no such thing is possible on the airless Moon, because liquid water could never have existed there. The rille was carved by flowing lava. The sinuous feature, a prime target for Apollo 15 mission, is named after the nearby Mount Hadley, the obviously tall peak at the top right of this view.

3 EARTHRISE

Apollo astronauts took many pictures of Earth rising above the lifeless lunar horizon. Most of us will have seen this, or some similar image, reinforcing our sense of Earth's fragility against the black and empty immensity of space.

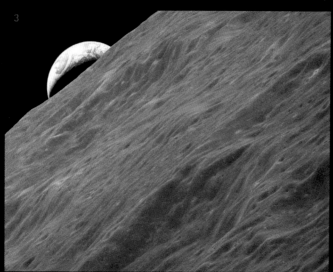

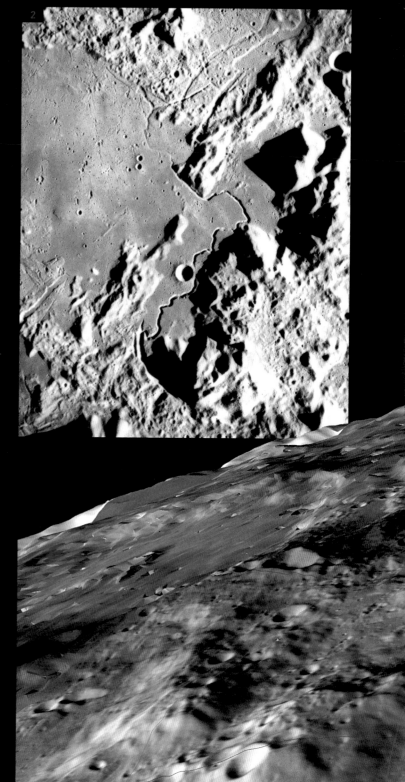

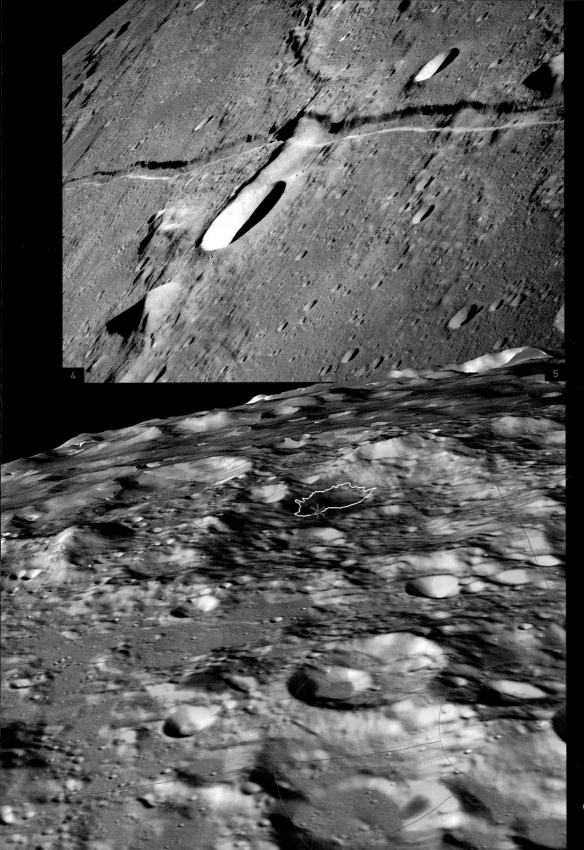

4 LUNAR RILLES

A close-up photo of the Aridaeus Rille, taken by the crew of Apollo 10, shows how the channel was formed by lava flowing underneath the lunar surface. Then the roof of the channel collapsed, leaving this characteristic valley-like structure.

5 WATER ON THE MOON

Even though no liquid water exists on the Moon, there is evidence of water ice nestling in the shadows of deep craters, or buried under the soil. The water was probably brought to the Moon by cometary impacts. Future lunar explorers may be able to extract oxygen and hydrogen from the water ice. The blue areas show terrain rich in hydrogen, probably bound up in frozen water molecules. In October 2009 a stage of the NASA's Lunar CRater Observation and Sensing Satellite (LCROSS) spacecraft was deliberately crashed into the permanently shadowed Cabeus crater near the South Pole, generating a plume of soil debris that could be analysed by other orbiting instruments. The red star marks the impact site, and the white digital graphic marker at bottom right indicates the polar axis.

THE RED PLANET

The 'Red Planet' has preoccupied us for more than a century. Fantasy writers and scientists alike have always speculated about the possibility of life on this world. Even today, we hope to find its microbial traces.

Mars has some of the most dramatic terrain in the Solar System, including four colossal volcanoes. The greatest of them, suitably christened Olympus Mons, covers an area equivalent to Arizona, and is capped by a crater that could swallow the entire island of Hawaii. Another prominent feature, Valles Marineris, is a tangled network of wide, deep canyons wrapping its way around half the planet. These landscapes are on such a grand scale, they can properly be appreciated only from space.

The largest Martian feature is, at first glance, the least obvious. The major volcanoes, the Marineris trenches, and large areas of rifts and fractures are associated with the Tharsis Bulge, a vast region where the planet's sphere was pushed out of shape by the internal pressures of molten rock beneath the crust. The volcanoes were fed by this great blister's hot interior. The magma's journey towards these volcanoes may have formed an underground channel whose crustal roof eventually fell in to create Marineris.

Scans taken from orbit reveal the presence on Mars of ancient river beds and sedimentary soils. Scientists still hope to find some microbial life underground, or at the very least, fossil traces from long-dead creatures. Mars did once have a much denser and warmer atmosphere than today, along with plenty of liquid water. Unfortunately, over billions of years, much of the atmosphere has escaped into space because of the planet's weak gravitational field, leaving a dry and frozen world soaked in harmful ultraviolet radiation from the Sun. The volcanoes, too, are no longer thundering.

THE CANYONS OF MARS *The largest canyon in the Solar System cuts a wide swath across the face of Mars. Named Valles Marineris, the system of rifts, channels and gulleys is more than 3,000 km (1,860 miles) long and up to 8 km (5 miles) deep. By comparison, the Earth's Grand Canyon in Arizona is the merest scratch. The origins of this dramatic scar are uncertain. The Martian crust may have cracked billions of years ago as the planet cooled.*

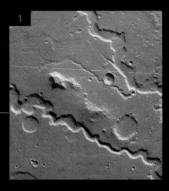

1 EVIDENCE FOR WATER

At some time in Mars's past, powerful flows of water carved many deep channels, such as these sinuous examples at Nanedi Vallis. The young planet must have had a thicker atmosphere and warmer climate than it does today.

2 FROZEN IN PLACE

This crater near the Martian North Pole, situated in the Vastitas Borealis plain, is 35 km (22 miles) in diameter. A huge lake of water ice is clearly visible, stranded in magnificent isolation after the waters that originally fed into the crater retreated.

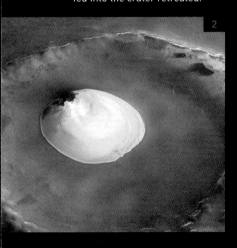

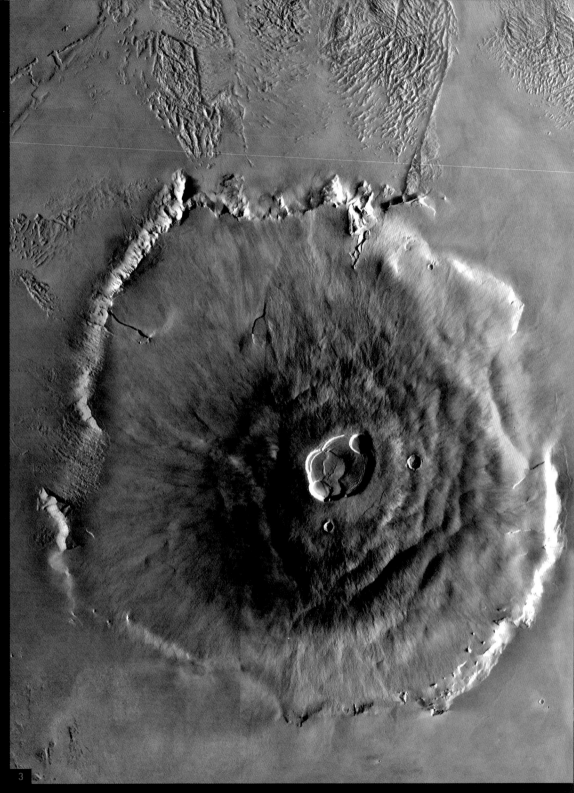

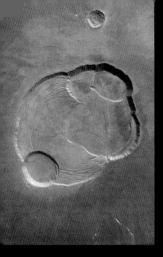

5 Southern ice cap

The southern polar ice cap, seen here during the Martian summer months, is made from a mix of carbon dioxide and water ices. In winter the ice cap doubles the extent of its surface coverage.

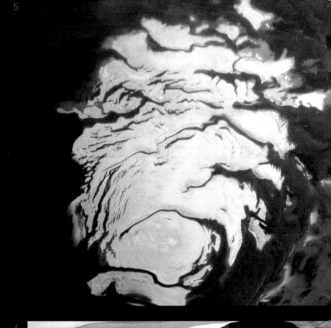

Multiple eruptions

A more detailed view of Olympus Mons shows overlapping vents, or calderas, resulting from a series of eruptions. The smaller calderas are the youngest. Obviously the volcano diminished in explosive power over time. It is unlikely to blow up again.

6 North pole

The northern ice cap consists of water ice overlaid with a thin frost of carbon dioxide. This perspective view shows distinctive stepped terraces. Similar features are found at both of the Martian poles. The terraces may have formed as a result of wind erosion.

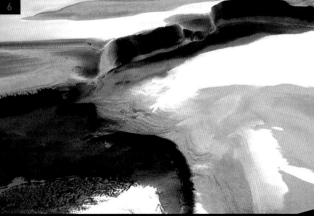

3 Vast volcano

Olympus Mons is the largest volcano in the Solar System. Although it is three times taller than the Earth's highest peak, Mount Everest, Olympus Mons would not be difficult to climb, because its slopes ascend very gradually over an area wider than the entire Hawaiian volcano chain. The low gravity of Mars, combined with a relatively static surface crust, allowed the volcano to build up over time, during multiple eruptions

7 Layers of ice

Polar ice pushing forwards and retreating with the changing seasons leaves its mark on the underlying terrain. Closer scans of layers at Chasma Boreale, a deep canyon near the North Pole show ice-rich tiers deposited between darker beds of dust. As with so many Martian features, the exact mechanisms that created these landscapes are not yet fully understood.

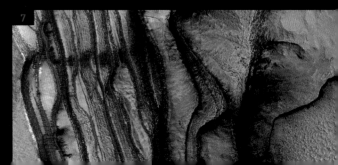

Not canals

This view of the so-called Schiaparelli hemisphere of Mars is named in honour of the Italian astronomer who, in the late 19th century, thought he could see what he called 'canali' on Mars. He just meant 'channels', but his work was interpreted by other astronomers to mean 'canals', artificial water systems built by intelligent beings. Today we know that varying patches of light and dark terrain on Mars, seen through low-powered telescopes on Earth, created misleading illusions.

2 Martian landscape

Fine red soil peppered with countless volcanic rocks typifies the Martian terrain, especially in the flat plains favoured as safe targets for robot landers. The redness in the Red Planet is a product of its thin topsoil, which contains rust-coloured iron oxides.

2 →

3, 4, 5 MACHINES ON MARS

Several landers are sitting on the Martian surface, well preserved in the cold, arid conditions. A heat shield discarded by Mars Rover Opportunity [3] is visible at the top left, and at right we see the crater gouged out when the shield first hit the ground. Opportunity's twin, the Spirit rover, landed in Gusev Crater, a region thought to have held a lake at some time in Mars's past. These rocks [4] look as if they were made by sediments deposited by water, but initial visual clues can be misleading. These were more likely volcanic in origin. A more subtle clue is in the sky [5] where a constant, gentle pink glow reveals the presence of fine dust in the Martian atmosphere. That dust may contain clay minerals. Clays are made by slow erosion of rocks by water.

6 FOREIGN OBJECT

This rock may be resting on the Martian surface, but it is not part of the planet's native material. It is a nickel-iron meteorite, here being examined up close by NASA's Opportunity rover in September 2010.

1

1, 2 DUSTY WORLD

Space probes have occasionally arrived at Mars only to find global dust storms obscuring all the surface features they are attempting to study. When a storm is raging, only the peaks of the tall volcanos can be seen from space [1]. As the dust retreats, the rich details of the surface are slowly revealed [2]. The orbit of Mars around the Sun is not quite circular. Its distance from the primary source of warmth varies according to the seasons, and these shifts affect its weather. Global dust storms tend to occur during perihelion, the period of closest approach to the Sun.

2

3 MARINERIS DETAIL

This central region of the Valles Marineris system, known as Ophir Chasma, is a chaotic landscape of trenches and plateaus. Narrower channels have collapsed into the deeper valleys below. These falls may have been triggered by ancient flooding and mudslides from sudden releases of meltwater.

4, 5 MARTIAN MYSTERIES

Relatively young craters pepper the landscape of Mars. This strange chain of craters at Phlegethon Catena [4] has not yet been fully explained. They may be the result of subsidence rather than impact. Some equally strange-looking rivulet shapes on the flanks of the Pavonis Mons volcano [5] are probably the scars of lava flows.

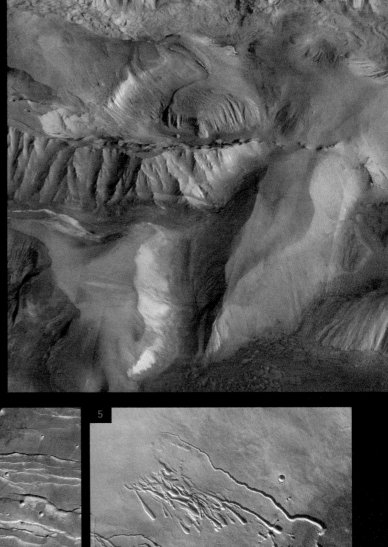

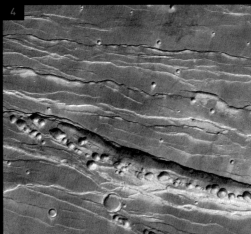

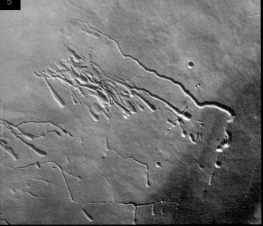

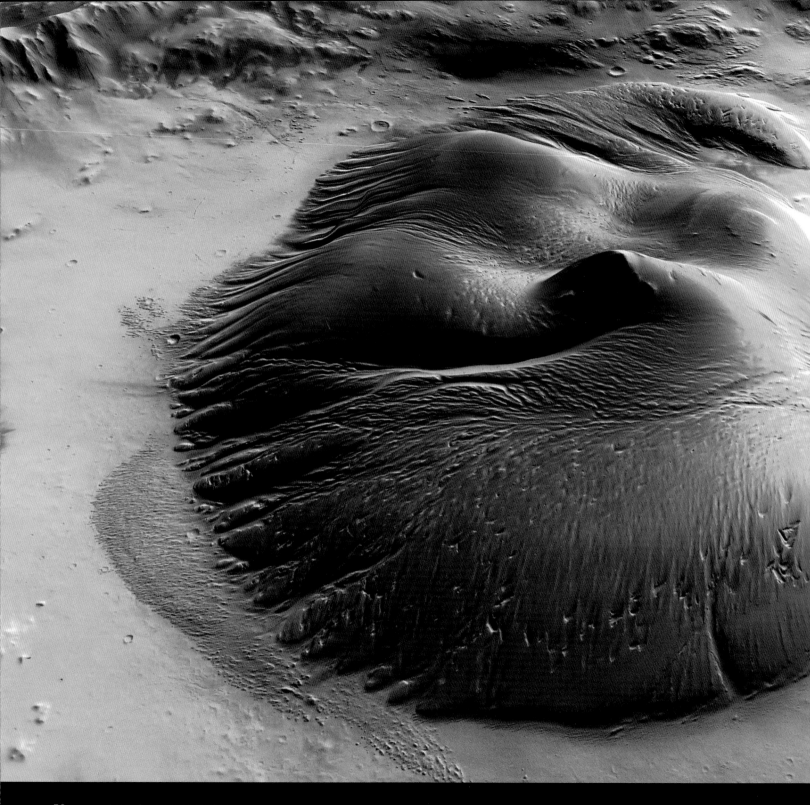

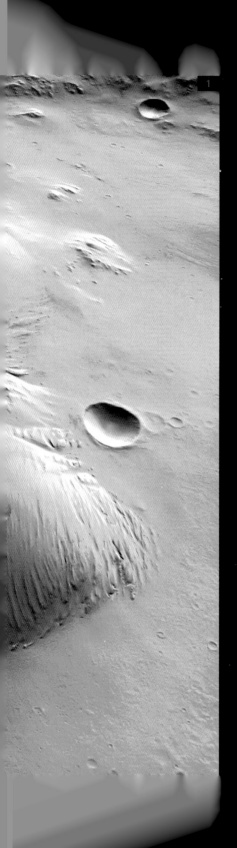

1 Domed crater

The floor of the 100-km diameter (62-mile) Nicholson Crater is raised in the middle. This is a common feature of impact craters, created as the distorted crust settles back after the initial shock and heat of the smash. Grooves and channels indicate that some kind of erosion processes have worn away at the dome-like central structure.

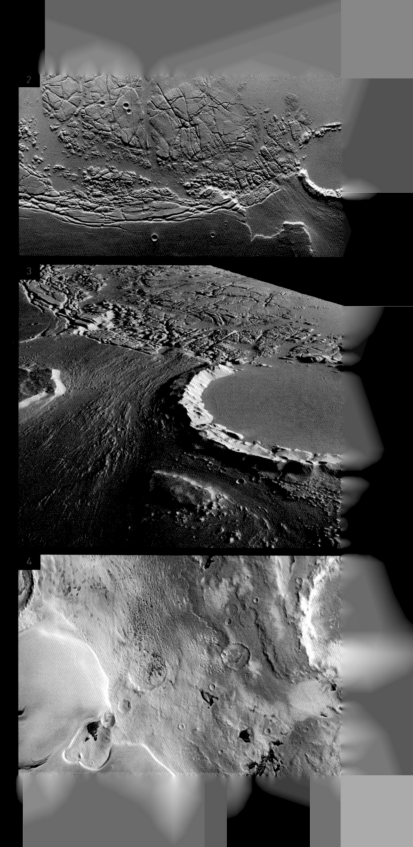

2, 3, 4 Variable terrain

An overhead view of the Kasei Vallis and Sacra Foisae regions [2] show chaotic disruption, possibly caused by brief and extremely violent flooding. A perspective rendering of radar data shows that the sides of the large crater in that region [3] appear to have been eroded by wind and water. In its liquid guise, that water has long since vanished, but this shot of Promethei Planum [4] shows a layer of water ice as deep as our Mediterranean Sea.

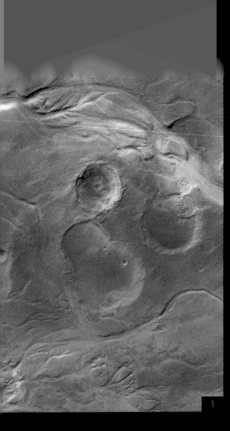

1

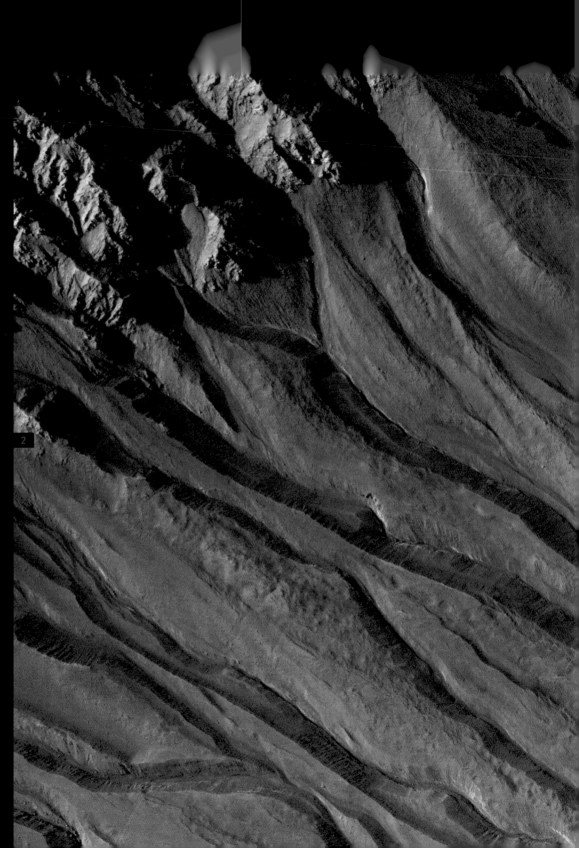

2

1, 2 Carved by water?

This outflow channel at Mangala Vallis [1] must have been cut by large quantities of water. Scientists are trying to work out if those great volumes surged just briefly in the past, or worked more slowly, over longer time spans. These sinuous gullies on the flanks of a southern crater [2] are suggestive of very recent water activity. Did liquid water exist for long enough to encourage the possible emergence of life, or are these gulleys just the product of wind and dry dust?

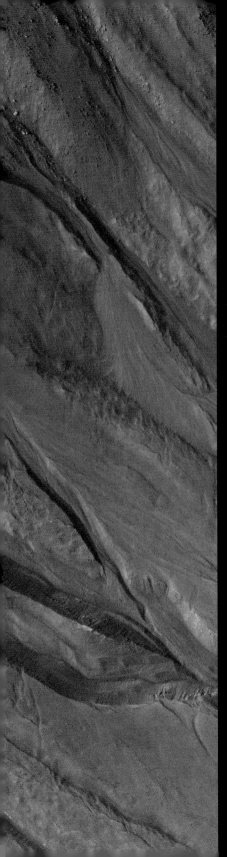

3 Martian minerals

This false colour image, based on data from the Mars Global Surveyor spacecraft, plots the distribution of various mineral types, including clays, at Mawrth Vallis, an outflow channel in the northern highlands. Iron and magnesium-rich silicate minerals are shown in red, and aluminium-rich silicates in blue, with other silica-based compounds in yellow and green.

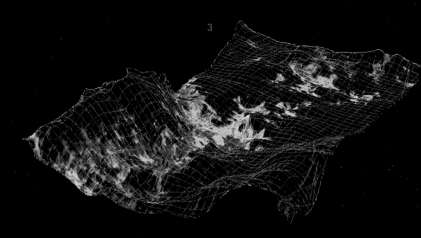

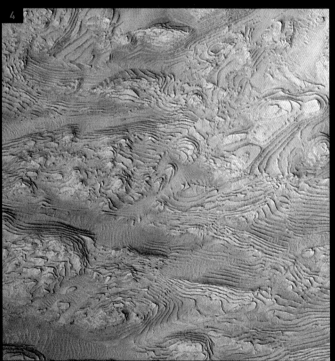

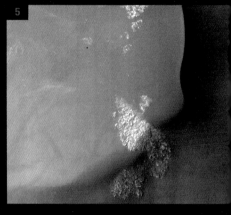

5 Dust sculptures

This crater in the southern hemisphere is filled with dark swaths of dust dunes. Different orientations show that the prevailing wind patterns have changed over time. The dark streaks crisscrossing the dune field are made by dust devils, low-powered whirlwinds that pick up dust particles and lay them down again elsewhere.

4 Layer after layer

Hundreds of layers of similar thickness, texture, and pattern have been exposed by erosion in a 64 km-wide (40 mile-wide) impact crater in western Arabia Terra. The layers are a record of repeated, episodic changes that took place at some time far in the martian past, as layers of sediment were deposited one on top of the other, time and again. They may have formed by dust settling out of the atmosphere, or perhaps by water-borne erosion products.

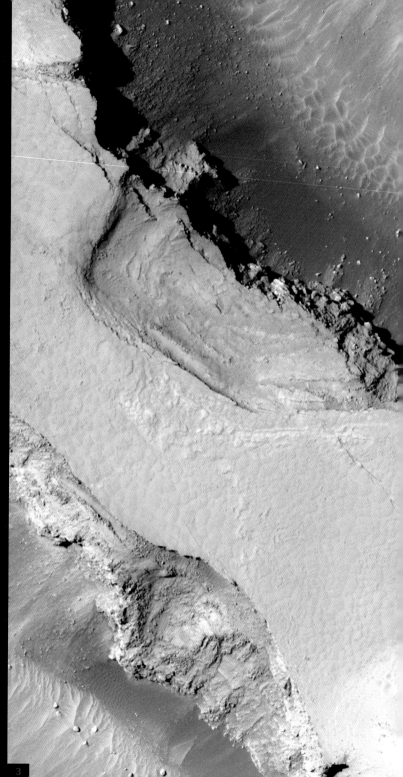

1 BREACHING THE WALLS

The bright, layered rock in this detail of Holden Crater is thought to have been deposited by a powerful flood that broke through the crater's rim.

2 SMASHED BY A METEORITE

Part of the floor of a large impact crater in the southern highlands, north of the giant Hellas impact basin. The fractures appear dark, probably as a result of wind-blown sand trapped in the cracks.

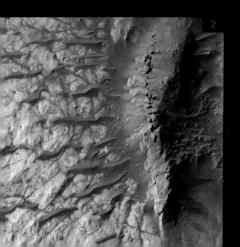

3 DELICATE DETAILS

A detail of Cerberus Fossae, a long system of faults in the crust. Cerberus Fossae was the source of a large volcanic eruption that draped an adjacent region, Athabasca Valles, in lava. The ripples in the middle of the deep channel are patterns of fine dust shaped by the wind.

1 Shadow of a visitor

Opportunity and its sister robot Spirit have been on Mars since early 2004. Here, Opportunity looks into Endurance Crater and sees its own shadow. Two wheels are visible on the lower left and right, while the floor and walls of the unusual crater are visible in the background.

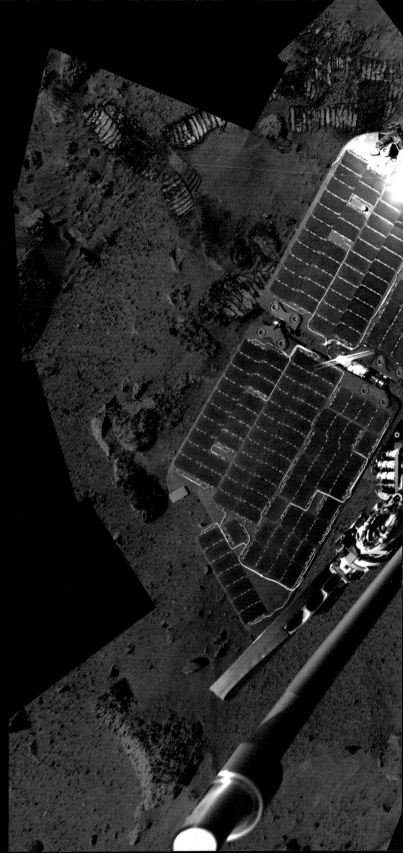

2 After arrival

As each rover descends from the safe haven of its carrier craft, it looks back and photographs the system of air bags and folding panels that delivered it onto the Martian surface.

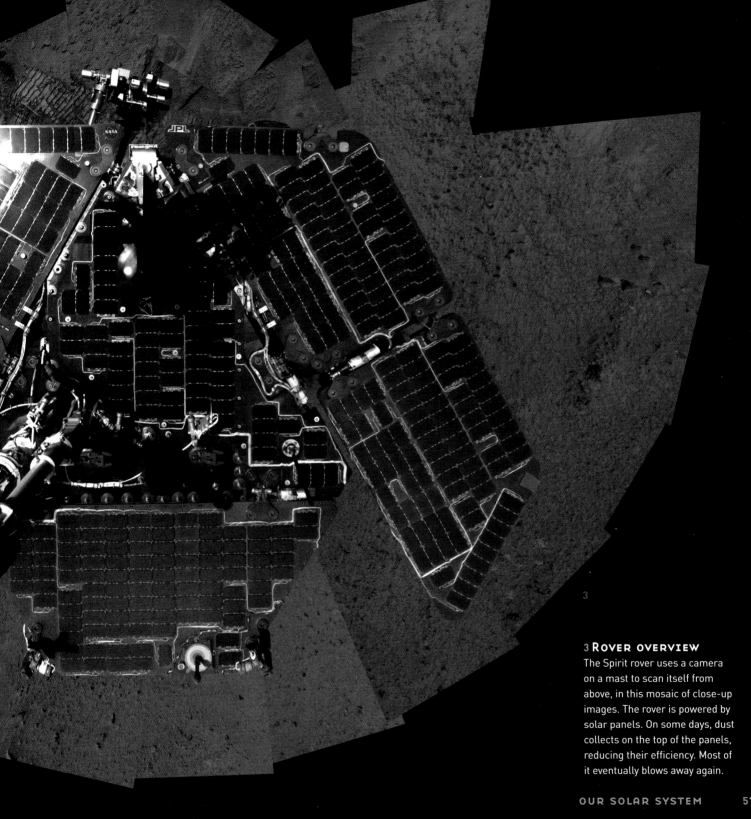

3 ROVER OVERVIEW
The Spirit rover uses a camera on a mast to scan itself from above, in this mosaic of close-up images. The rover is powered by solar panels. On some days, dust collects on the top of the panels, reducing their efficiency. Most of it eventually blows away again.

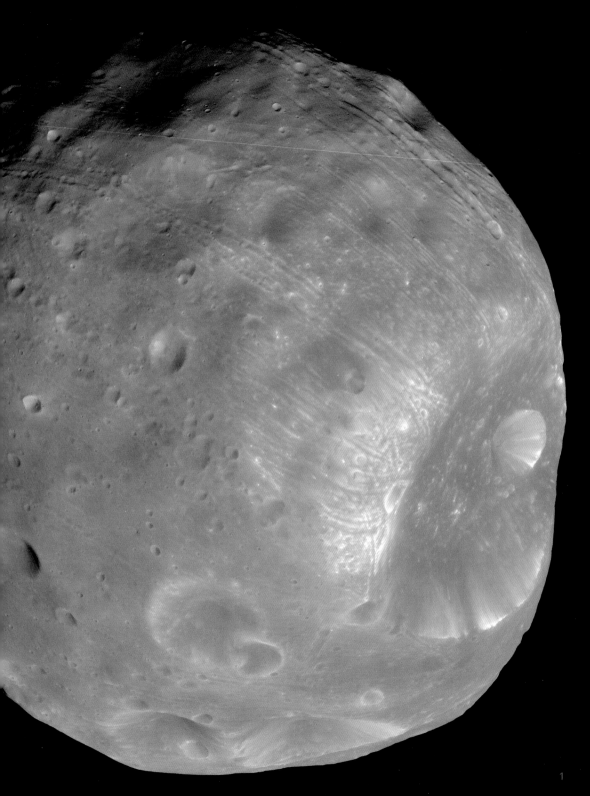

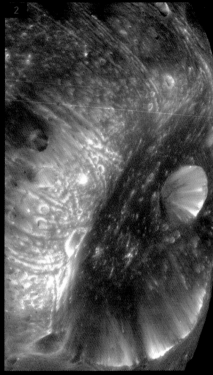

1, 2 PHOBOS CLOSE-UP

The Mars Reconnaissance Orbiter took this image of Phobos [1] in March 2008, from a distance of 6,800 km (4,200 miles). The most prominent feature is a large crater, called Stickney, at the lower right. A series of troughs and crater chains is obvious on other parts of the Moon. They may have formed when material ejected from impacts on Mars later collided with Phobos. A detail of Stickney Crater [2], which was named for Chloe Angeline Stickney, mathematician and wife of astronomer Asaph Hall, the man who discovered Phobos and Deimos in 1877.

3

4 VICTORIA CRATER

At 730 m (2,400 ft) wide, Victoria crater is a relatively insignificant feature in the landscape of the Meridiani Planum region close to the Martian equator. However, it is one of the most intensively studied regions of the red planet's surface thanks to its location close to the landing site of the NASA's Opportunity Mars Rover.

3 STRANGE PLATES

These blocks appear to be ice floating on a recently frozen sea, all of which is now covered by fine orange-red dust. The unusual plates were photographed by the European Space Agency's Mars Express spacecraft. Oddly, this region lies near the Martian equator and not near either of Mars's frozen polar caps, where major ice formations might be expected. Without the covering of dust, any water or ice in the equatorial regions would quickly evaporate. These plates are very similar in appearance to ice blocks in the seas around Antarctica.

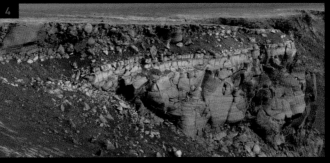

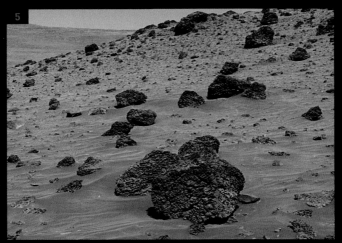

5 ODD ROCK

What created this unusually textured rock on Mars? Most probably, a volcano. Dubbed Bumpy Boulder, the strange stone, was found by the robotic Spirit rover. Pits on the ragged surface may have formed from gas bubbling out of hot rock ejected by a Martian volcano. Several similar rocks are visible nearby.

6 MARTIAN SUNSET

Although the atmosphere on Mars is extremely thin, a suspension of very fine dust particles diffuses the sunlight, especially when the Sun is low on the horizon. This mosaic image was taken by the Mars Pathfinder lander at Ares Valles during July 1997.

6

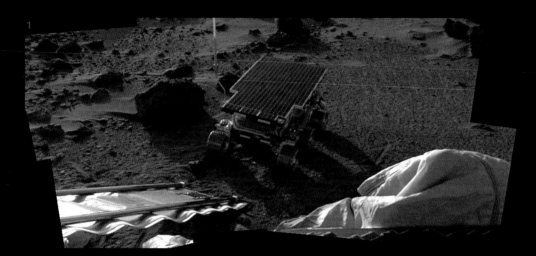

4 VICTORIA CRATER

This impressive feature at Meridiani Planum, here viewed by the Mars Reconnaissance Orbiter, is approximately 800 m (half a mile) in diameter. It has a distinctive scalloped shape to its rim, caused by erosion and downhill movement of loose material in its walls. Layered sedimentary rocks are exposed along the inner walls. The rippled features in the middle are products of wind-sculpted dust.

1 SETTING OFF

Sojourner, the first wheeled rover to explore the surface of Mars, moves away from the landing ramp of the Pathfinder delivery vehicle in July 1997. Part of the airbag landing system can be seen at lower right.

2 YOGI ROCK

Sojourner pushes up against a rock, named Yogi, and makes contact using an x-ray spectrometer instrument. Yogi turned out to be very like common volcanic basalts found on Earth.

3 MINOR IMPACT

Nereus Crater is a small impact feature on Mars with a diameter of about 10 m (33 ft) located just south of the planet's equator, in the otherwise relatively smooth Meridiani Planum region. It was discovered by the Opportunity rover in September 2009.

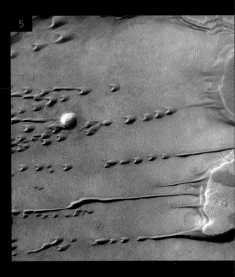

5 Liquid-like dunes

Although liquids freeze and evaporate quickly into the thin atmosphere of Mars, persistent winds can cause large sand dunes to flow and even drip as if they were liquid. The dark arc-shaped droplets of fine sand are called barchans, and are the interplanetary cousins of similar Earth-based sand forms. Barchans can move intact downwind and can even appear to pass through each other.

7 Darkest moon

Phobos, the largest and innermost of two Martian moons, is the darkest moon in the entire Solar System. Its unusual orbit and colour indicate that it may be a captured asteroid composed of a mixture of ice and dark rock. Phobos seems to be covered by a thick layer of loose dust. The tiny moon's orbit around Mars is continually decaying. It will likely break up and crash into Mars within the next 50 million years.

6 Moons of mars

Mars has two small moons, Phobos, shown here, and Deimos. They are only a few tens of kilometres in diameter, and their irregular shapes suggest that they must be stray asteroids captured by the Martian gravitational field. The grooves on Phobos have not been explained.

ASTEROIDS: THE WRECKAGE OF CREATION

In the 550 million kilometres (340 million miles) of space between the orbits of Mars and Jupiter lies the asteroid belt, where a vast mass of rubble circles the Sun in a diffuse and chaotic band.

More worlds than we see today condensed out of the Solar System's accretion disc, but at least a few of them developed orbits around the Sun that were too elliptical for safety. At some point in the distant past, their paths intersected and they smashed together, creating a swarm of rubble. Each chunk is known as an asteroid. Some asteroids may be the remains of an ancient planetesimal that was unlucky enough to stray too close to Jupiter. Tidal forces from the mighty gas giant tore the smaller world apart.

The Earth has been smashed into by rogue asteroids many times before. It is bound to happen again some time in the future. Meteorites are stray asteroids that collide with the Earth, survive their fiery descent and reach the ground intact. Some meteorites are made of iron, or a mix of stone and iron. Finally, the most common kind of meteorites are stony, very rich in silicate minerals, with only small flecks of metal. These are the wreckage from the outermost crusts of planetesimals.

CERES *The largest asteroid gained sufficient mass and gravity to pull itself into a spherical shape, classifying it as a 'dwarf planet'. Ceres is 960 km (596 miles) across, and orbits the Sun once every 4.6 years. Its terrain, viewed here by the Hubble Space Telescope, shows distinct variations in brightness.*

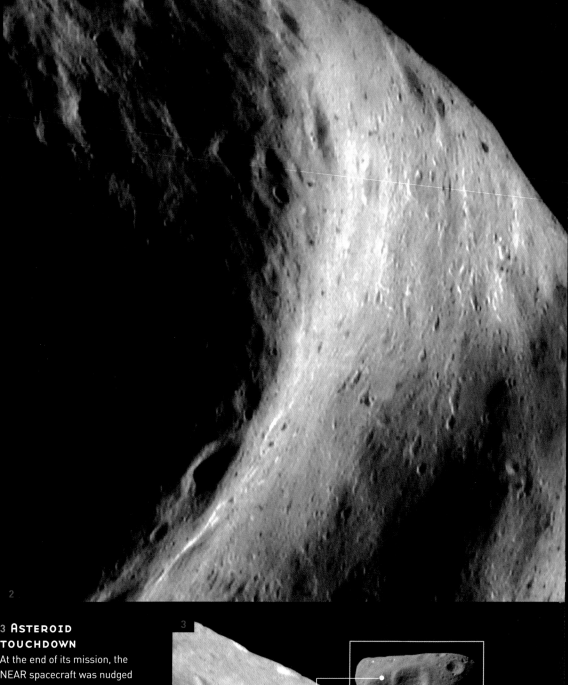

2

3 ASTEROID TOUCHDOWN

At the end of its mission, the NEAR spacecraft was nudged towards Eros for a deliberate, though very gentle, crash landing.

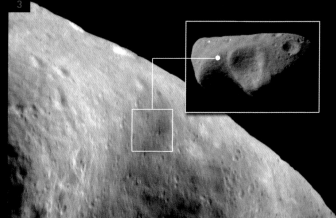

3

Dactyl

243 Ida

5535
Annefrank

9969 Braille

2867
Steins

433 Eros

1P/Halley

21 Lutetia

19P/Borrelly   9P/Tempel 1   81P/Wild 2

951 Gaspra

253 Mathilde

4 ASTEROIDS TO SCALE
A montage of all asteroids (and comets) visited to date by robotic spacecraft, but not including Mars's moons Phobos and Deimos, which are not yet proven to be asteroids. Five comets and ten asteroids are featured, although two of those asteroids, Ida and Dactyl, are gravitationally linked in a double asteroid system. Most of these targets were visited only briefly, in 'flyby' missions. Only Eros and Itokawa were orbited by their visiting probes, so that they could be photographed and mapped completely.

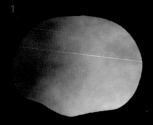

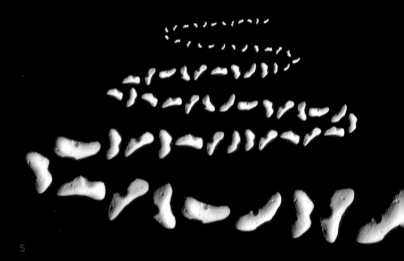

1 Failed sphere

At 560 km (348 miles) across, Vesta, the third largest asteroid, should have been large enough to pull itself into a spherical shape by virtue of its mass and gravity. This Hubble Telescope image shows that something must have prevented that from happening.

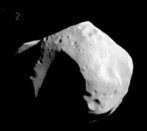

4, 5, 6 Eros approaching

This montage [5] shows a selection of images of asteroid 433 Eros acquired by the NEAR spacecraft over three weeks in early 2000, as the spacecraft closed in on its target. The asteroid's natural colour [4] is a subtle butterscotch hue. In June, 2000, NEAR trained its camera on Eros's largest crater [6]. The darker hues represent rock and regolith, loose dust and rubble, that has been altered chemically by long exposure to solar radiation. Lighter patches show areas apparently less affected.

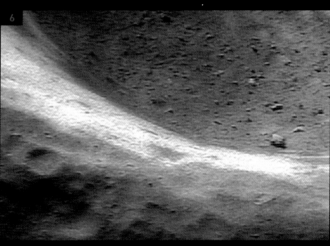

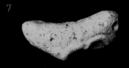

2 Mathilde

This asteroid is a fairly sizable 66 km (41 miles) in diameter, but is not as dense as might be expected from such a hefty lump of rock. Its gravity field is only as strong as the effect that a similar volume of water would exert. It may be peppered on the inside with cavities.

3 Gaspra

This small lump of rock is an S-type asteroid, rich in silicate materials and metals. In the future, asteroids such as Gaspra may make tempting targets for commercial exploitation.

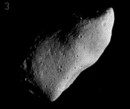

7, 8, 9 Strange gravity

Three views of Eros from the orbiting NEAR spacecraft [7, 8, 9] map the asteroid's varying gravitational pull as different colours overlapped on 3D scans of the asteroid created from laser rangefinder data. The asteroid's shape, density and spin combine to create a bizarre pattern of gravitational influences. For example, a ball dropped onto one of the red areas would try to roll 'uphill', then across the nearest green area and into the nearest blue area.

10 Dramatic light

A tremendous view of Lutetia, sidelit by the Sun, gives a menacing impression of this huge mass of rock.

10

12 Ancient survivor

The European probe Rosetta made its closest approach to asteroid 21 Lutetia on July 10, 2010, revealing a battered world with many craters. Lutetia is probably a primitive survivor from the violent birth of the Solar System, some 4.5 billion years ago.

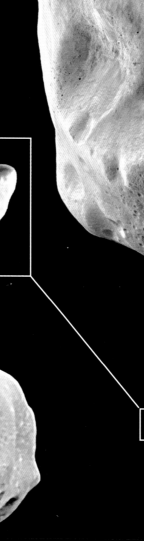

11 Ida and Dactyl

The Galileo probe discovered that Ida has its own miniature moon, a small, irregularly shaped lump of rock about 54 km (33.5 miles) long, and composed of carbon-rich materials.

11

1 LUTETIAN AVALANCHE

A detail of a crater-like feature on Lutetia shows that boulders have participated in a landslide, tumbling down the sides of the depression. The hollow was probably gouged by an impact, but its sides have become softer and more rounded than might be expected.

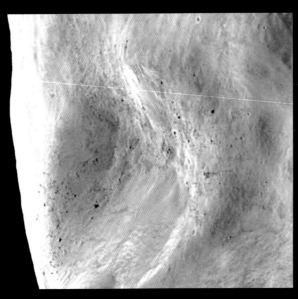

2, 3 ASTEROID COLLISION

In February 2010 the Hubble Space Telescope observed a mysterious X-shaped debris pattern [2] and trailing streamers of dust, suggesting a head-on collision between two asteroids. Astronomers have long believed that the asteroid belt is being gradually ground down through collisions, but such a smash has never been seen before. The primary object [3] was thought to be a rare example of a comet travelling near the asteroid belt. Later analysis told a more violent story.

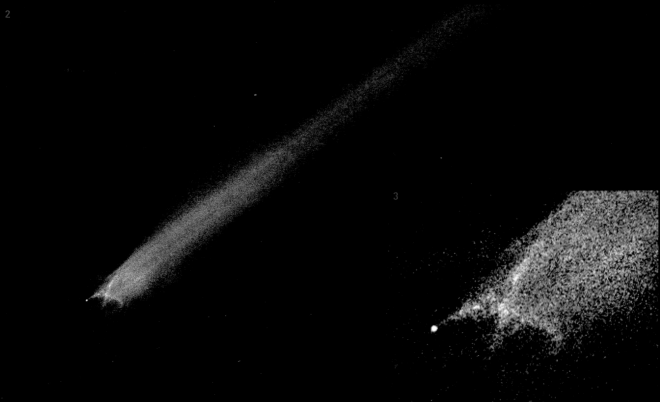

4, 5, 6, 7, 8, 9 JAPAN'S TOUCHDOWN

In September 2005 Japan's deep space probe Hayabusa arrived at asteroid Itokawa, 300 million km (186 million miles) away from the Earth [4, 5, 6, 7]. In November of that year, it successfully landed and collected small amounts of material for return to Earth. A detail of the touchdown zone and its location on the asteroid's flank is shown here [8]. The shadow of the Hyabusa probe is evident as it closes in for touchdown [9]. At the end of its visit, Hyabusa set off once again into deep space for the long journey home. In June 2010 the probe's reentry capsule came down in outback terrain at Woomera in Australia, and was recovered. A few tiny grains of Itokawa were inside the sample canister.

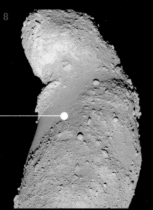

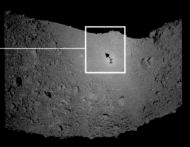

THE GAS GIANTS

What is the difference between a star and the planets that form around it? The Solar System's largest planet shows that the line separating one from the other can blur.

When the Solar System first began to form, the heat radiating out of the new Sun blew away the hydrogen molecules in the surrounding accretion disc, so that the regions closest to the Sun became, by default, richer in the scarce heavier chemical elements like iron and silicon, while the outer regions of the disc, where the Sun's radiation was not so intense, were able to hold on to the more widely available hydrogen. This explains why the inner planets of the Solar System are small, rocky and dense, while the outer worlds are mainly giant spheres of gas, dominated by hydrogen.

Jupiter, the greatest of the four Gas Giants. Given more time and gaseous material to feed upon, this massive hydrogen-rich ball might have become a star in its own right, a 'brown dwarf' too small to reach ignition point, but a kind of star nevertheless.

Saturn's famous ring system consists mostly of water ice and grains of dust. The rings extend hundreds of thousands of kilometres from the planet, yet their vertical depth is negligible. Viewed edge-on, they almost vanish. Uranus, the third gas giant, is a dynamic world with some of the brightest clouds in the outer solar system, and its own faint ring system. The atmosphere is mainly hydrogen and helium, with a small amount of methane. Neptune, a similar world to Uranus, orbits the Sun once every 165 years. Unlike all the other planets, it is invisible to the naked eye because of its extreme distance from the Earth.

BULGING GIANT *Jupiter is the fastest-spinning planet in the Solar System. It takes less than ten hours to complete a rotation around its axis. The rapid rotation of this massive planet causes its equator to bulge outwards. Rotational energy also shapes the atmosphere, the clouds and the storm systems into distinct bands and whorls. The famous Great Red Spot at the lower centre of this image was first observed by astronomers more than 300 years ago, and shows no signs of abating today.*

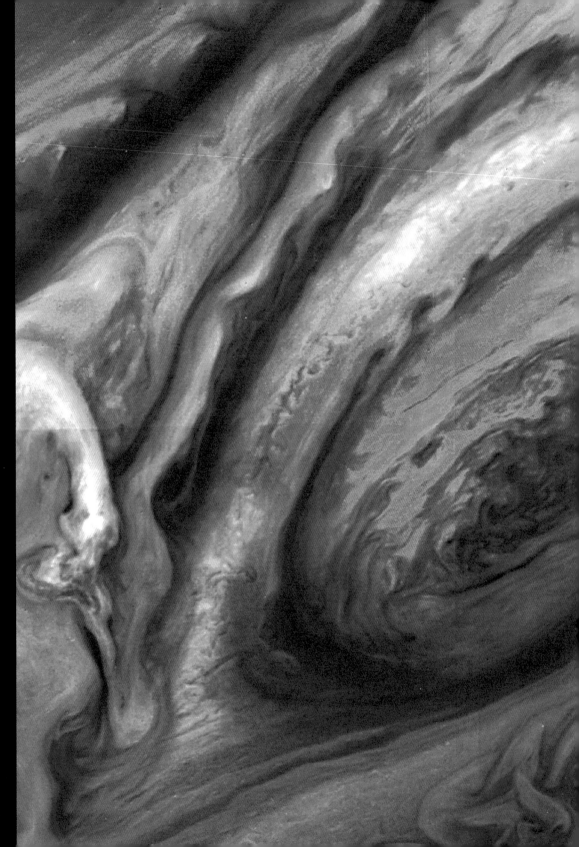

1

1 FAMILIAR GIANT
Jupiter's Great Red Spot and
the banded features of the
atmosphere can be seen in
any reasonably decent optical
telescope, but professional
astronomers tend to augment
the details through multispectral
analysis.

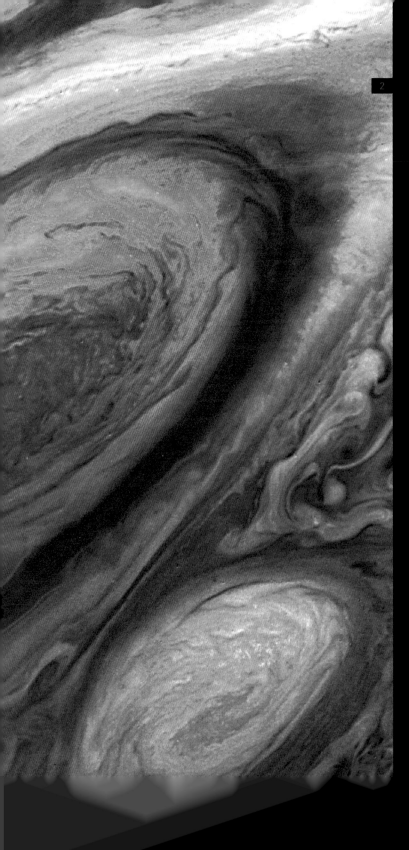

2, 3, 4, 5 THE ETERNAL STORM

No one knows how long Jupiter's famous Great Red Spot storm system [2] has raged, nor how long it will continue. It seems to behave like a hurricane on Earth, but it is enormous. Three Earths would fit within its boundaries. Located in Jupiter's Southern hemisphere, its rotation is counterclockwise, with a period of about six days. This succession of images [3, 4, 5] show the Red Spot moving through the upper atmosphere.

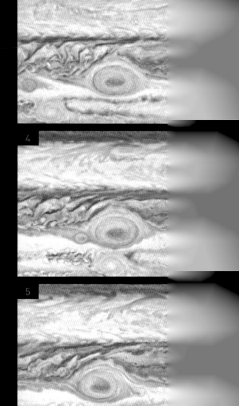

6 MAP OF THE WEATHER

This is how Jupiter looked on 12 December 2000. Small bright spots in the dark bands above the equatorial region are thunderstorms.

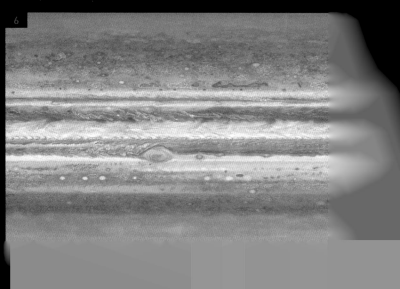

1 Polar view

We never see this view of Jupiter through our telescopes, but this image made up from narrow angle images taken from the Cassini spacecraft, is what the familiar planet looks if like if seen from above its North Pole.

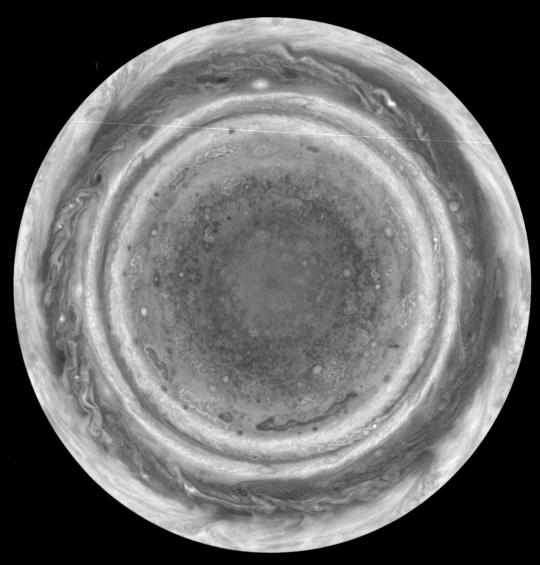

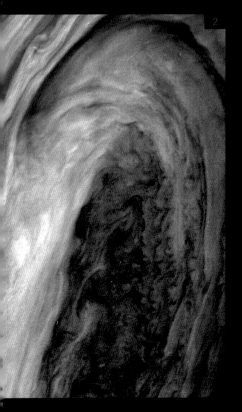

2 Into the storm

A close-up image of the Great Red Spot, revealing the complexity of its structures and energies.

3, 4 No less dramatic

Time-lapse images of other weather systems and atmospheric features on Jupiter. If the Great Red Spot did not exist, the rest of the planet's weather systems would still impress us with their immense scale.

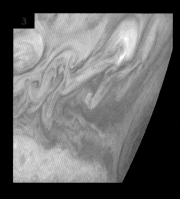
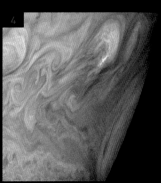

5 Family portrait

This composite of the Jovian system's major objects includes the edge of Jupiter with its Great Red Spot, and Jupiter's four largest moons, known as the Galilean satellites in honour of Galileo, the astronomer who first discovered them. From top to bottom the moons shown are Io, Europa, Ganymede and Callisto. The moons are to scale, but are artificially grouped together for clarity.

6 The rings of Jupiter

Saturn's rings are so obvious that they can easily be seen in any decent telescope. Jupiter's rings are more subtle, but backlighting from the Sun reveals them. The rings were only discovered in 1979, when the first space probes visited the planet.

1 ACTIVE IO

Two simultaneous eruptions are visible on this image of Io viewed from the Galileo probe. On the right, a bluish plume rises above the surface of a volcanic caldera known as Pillan Patera.

In the middle of the image, near the shadow line where night meets day, another plume is seen from directly overhead, casting a dark shadow to the left of the volcanic vent that ejected it.

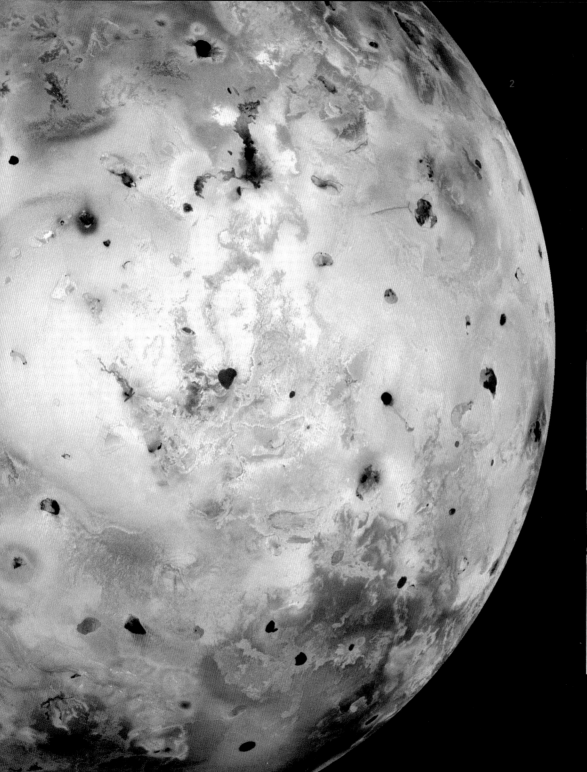

2, 3, 4 IRRITABLE IO

Sulphurous yellow compounds
stain the surface of Io [2]. Red
areas seem to indicate fresh
lava flow. Io is affected by
immense tidal forces from
Jupiter. As a consequence,
the moon's crust is never stable,
and its molten interior constantly
breaks through onto the surface.
Details of Io's surface [3,4] show
hot molten lava bursting through
the obviously fragile crust.

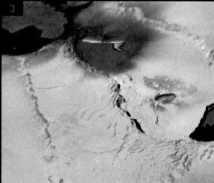

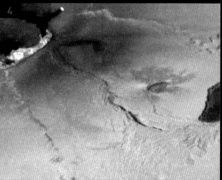

4 IO IN TRANSIT

Jupiter's innermost moon crosses the face of its incredibly huge parent planet in this haunting image from the Cassini space probe.

5 JUPITER AND IO →

NASA's Pluto-bound New Horizons spacecraft passed through the Jupiter system in early 2007, riding the planet's gravity to boost its speed and shave three years off its trip to Pluto. As it passed, it captured this spectacular view of Io in the foreground.

1 IO'S SHADOW

Another extraordinary view of Io as it passes across (transits) the face of Jupiter, casting a deep black shadow on the top of the Jovian atmosphere.

2, 3 IO ECLIPSE

When Io interposes itself between a spacecraft and the Sun, the eclipse enables the probe to monitor delicate phenomena that would otherwise be swamped by solar radiation. These images [2, 3] show varying accumulations of volcanic gases in Io's thin atmosphere.

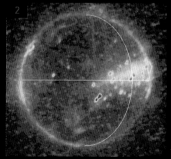

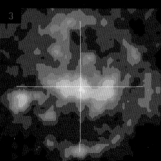

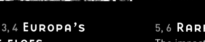

1, 2, 3, 4 EUROPA'S ICE FLOES

Detailed views of the Europan surface reveal features that look almost exactly like pack ice on a terrestrial arctic sea [1]. The floes break up, then refreeze in new positions. An even closer look at cracks between the floes [2, 3, 4] shows brownish material that seems to have oozed up between the cracks, then frozen in place. The brown discolouration may come from the subsurface water. Scientists think that this water may contain organic materials.

5, 6 RARE CRATER

The impact crater Pwyll is named for a character in Celtic mythology. Pwyll's visible dark central region is about 26 km (16 miles) in diameter, while brilliant white rays of debris blasted from the impact site extend outward for hundreds of kilometres. The white debris clearly overlays everything else on the Europan surface, indicating that this crater is younger than the surrounding features.

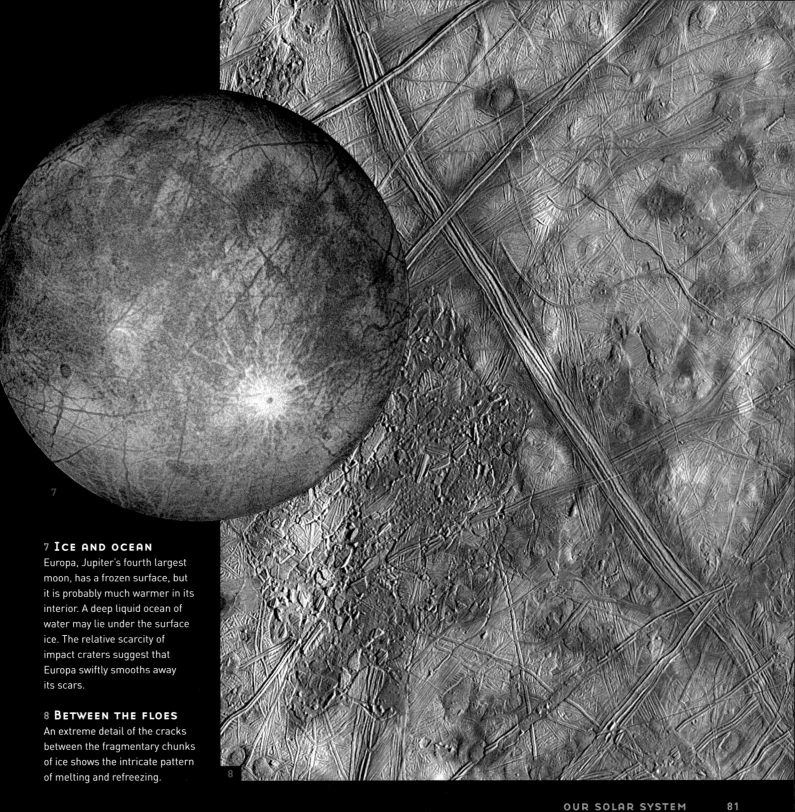

7 Ice and ocean

Europa, Jupiter's fourth largest
moon, has a frozen surface, but
it is probably much warmer in its
interior. A deep liquid ocean of
water may lie under the surface
ice. The relative scarcity of
impact craters suggest that
Europa swiftly smooths away
its scars.

8 Between the floes

An extreme detail of the cracks
between the fragmentary chunks
of ice shows the intricate pattern
of melting and refreezing.

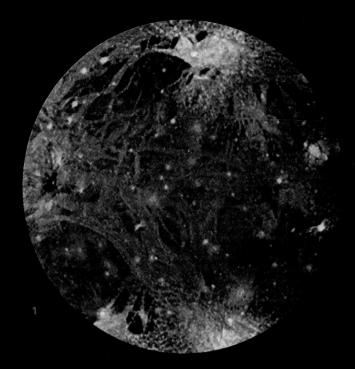

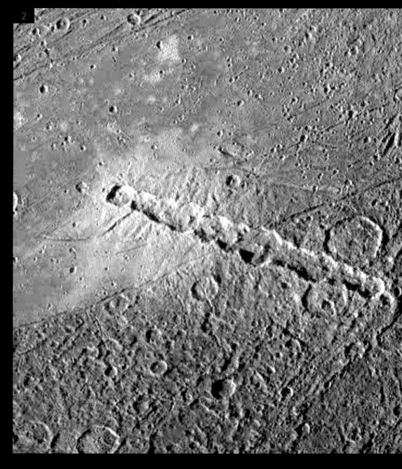

1 Largest satellite

Ganymede is Jupiter's largest moon, and the largest in the entire Solar System. It is also the only moon with its own magnetic field. This suggests that it has a molten iron-rich core.

2 Multiple hits

A chain of craters on Ganymede, named Enki Catena, may have been created by a comet that was pulled into pieces by Jupiter's gravity as it passed too close to the planet. Soon after this breakup, the fragments crashed onto Ganymede in rapid succession.

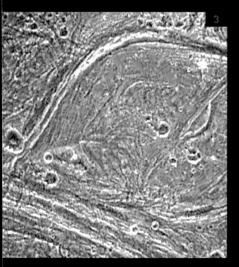

3, 4 Shifting plates

Complex tecton crustal plates are evident in this image of Ganymede's surface, showing the border between a region of ancient dark terrain known as Marius Regio, and another area of younger terrain, named Nippur Sulcus [3]. The curved appearance of this feature is probably due to simultaneous twisting and sliding of the plates. A computer model of Ganymede's surface [4] shows its deep grooves caused by tectonic distortions in the crust.

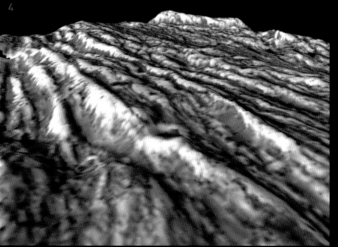

5 Brightest hit

The light branching bands from the most obvious impact crater on Ganymede suggest that a thick underlay of ice has been exposed by the force of impact.

6 Lightweight world

A distant view of Ganymede, which is about 1.5 times the size of our Moon, but only half as dense. It is probably composed of a mixture of rock and ice.

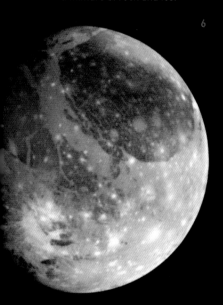

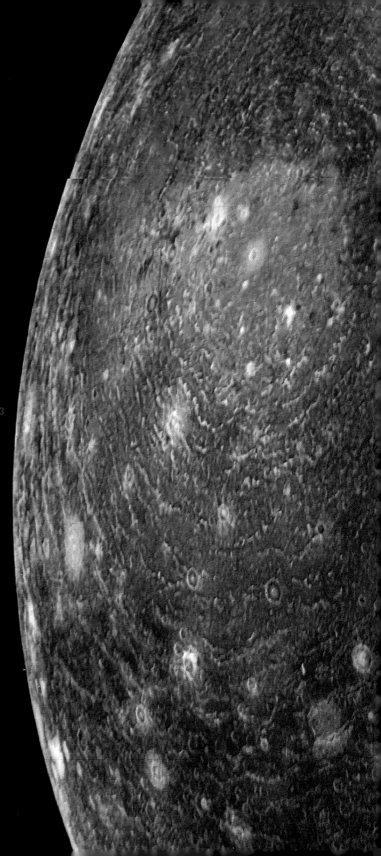

1, 2 CALLISTO

This picture of Callisto [1] was taken by Voyager 2 from a distance of 2,318,000 km (1,438,000 miles). Callisto is covered with bright spots, which are meteorite impact craters. The largest of them, a prominent bullseye-shaped feature [2], is similar to Mare Oriental on our Moon, or the Caloris Basin on Mercury.

3 ANCIENT REMNANT

Jupiter's second-largest moon is about the same size as the planet Mercury. It orbits just beyond the reach of Jupiter's fierce radiation belt. Callisto is the most heavily cratered satellite in the solar system. Its crust is very ancient and dates back to just shortly after the solar system was formed.

3

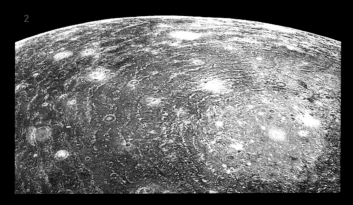

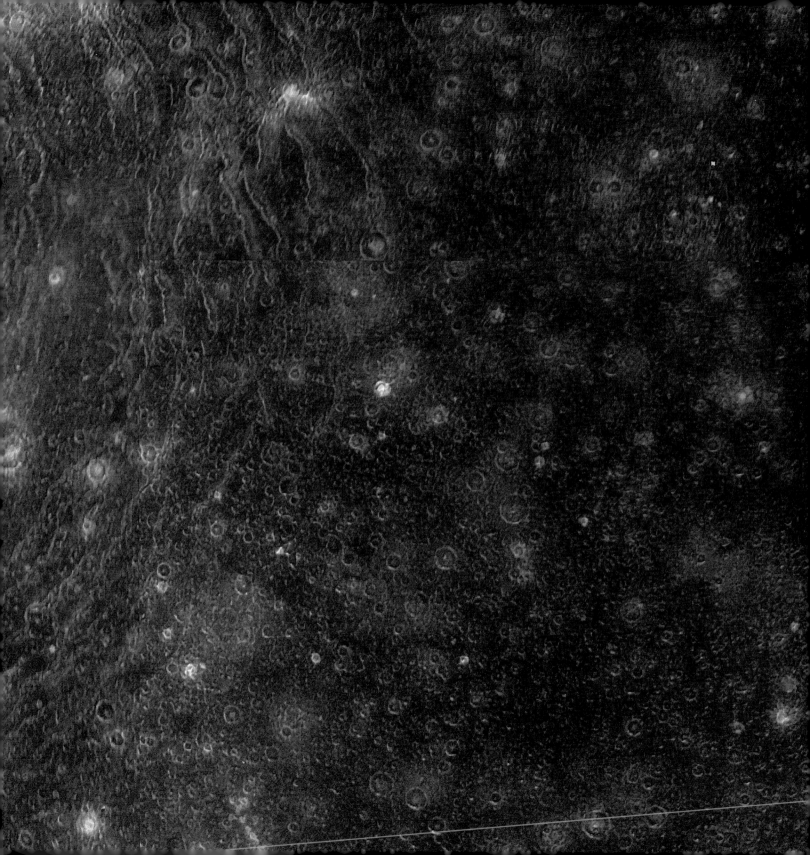

1, 2 BEAUTIFUL GIANT

Saturn is the second largest planet in the Solar System [1] with an equatorial diameter of 119,300 km (74,130 miles). It is visibly flattened at the poles, a result of the very fast rotation of the planet on its axis. Its day is 10 hours, 39 minutes long, and it takes 29.5 Earth years to orbit the Sun. The atmosphere is primarily composed of hydrogen with small amounts of helium and methane. The planet as a whole is less dense than water. In the unlikely event that a spectacularly vast ocean became available, Saturn would float in it like a cork. The distinctive and beautiful rings [2] are split into different zones. The most notable gap is the Cassini Division, the wider of the two dark bands clearly visible here.

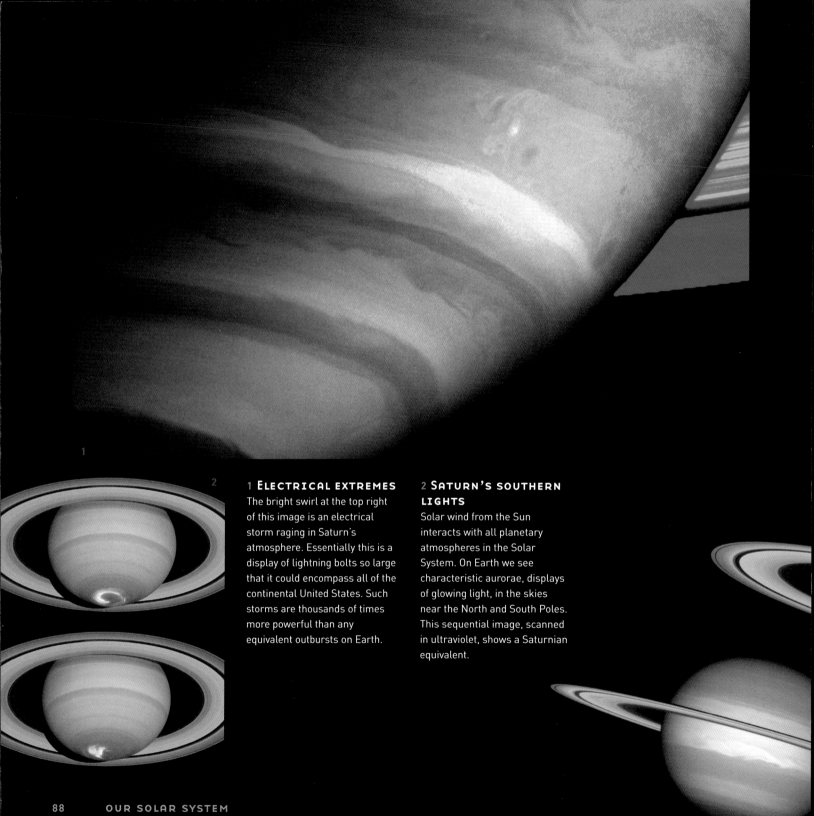

1 Electrical extremes

The bright swirl at the top right of this image is an electrical storm raging in Saturn's atmosphere. Essentially this is a display of lightning bolts so large that it could encompass all of the continental United States. Such storms are thousands of times more powerful than any equivalent outbursts on Earth.

2 Saturn's southern lights

Solar wind from the Sun interacts with all planetary atmospheres in the Solar System. On Earth we see characteristic aurorae, displays of glowing light, in the skies near the North and South Poles. This sequential image, scanned in ultraviolet, shows a Saturnian equivalent.

3 DANCE OF THE RINGS

This sequence of five images, captured at regular intervals by the Hubble Space Telescope between 1996 and 2000, show how our view of Saturn's ring system changes over time.

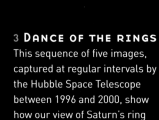

4 STUNNING SATURN

The Cassini spacecraft drifted in Saturn's shadow for about twelve hours and looked back toward the eclipsed Sun. The night side of Saturn is partly lit by light reflected from its ring system. Visible in spectacular detail is the outermost ring, created by the newly discovered ice-fountains of the moon Enceladus.

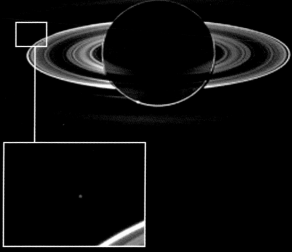

1 LINE AND SHADOW
The rings of Saturn seen edge-on vanish to the thinnest line, while casting vast shadows across the top of the planet's atmosphere. Moons appear as tiny 'bumps' in the ring line.

2, 3 EPIC SHADOWS
Seen from high above Saturn's North pole, the planet's bulk [3] casts a colossal shadow over its rings. An edge-on view obtained by Cassini in March 2006 shows the thin sliver of the rings silhouetted against Saturn with Rhea hanging below the unlit rings [2].

4 Tiny earth

As the beautiful backlight conditions change from Cassini's point of view, and the glare fades, Earth becomes visible as a distant and tiny dot of light in the background.

5 Real colours

Most images returned by space probes and orbiting telescopes are colour-enhanced to pick out subtle details. This is how Saturn looks when seen through a conventional telescope, and through human eyes. It is still intensely beautiful, but special instruments are required to see its underlying atmospheric dramas.

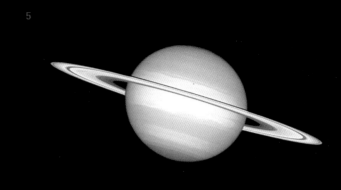

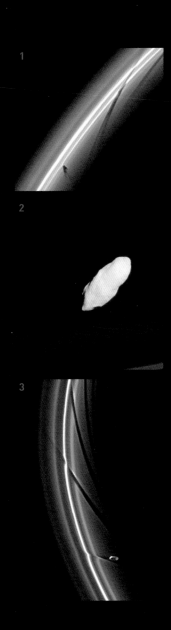

1, 2, 3 PUNCTURING THE RINGS

The tiny moon Prometheus passes upwards through Saturn's 'F' ring [1] and eventually pulls clear. What looks like a shadow behind Prometheus [2] is in fact the path of clear space that its gravity has temporarily gouged through the ring. The influences of moons may be at least part of the cause of spokes and other patterns [3] within the rings. The mechanisms that drive these complex and shifting formations are not yet understood.

5 SATURN IN FALSE COLOURS

The composite image was made from 65 individual observations by Cassini's visual and infrared mapping spectrometer in the near-infrared portion of the light spectrum.

6 STRANGE SPOKES

Spokes, kinks and distortions in the ring system form and dissipate over a few hours. They may be localized sheets of small dust-sized particles, shaped into patterns by Saturn's electromagnetic activity, and perhaps also by bursts of electric charge generated by lighting storms.

4 RING IMPERFECTIONS

Scientists once thought that Saturn's rings were almost completely flat, and extremely regular in construction, but new images reveal occasional strange bumps, some as high as the Rocky Mountains.

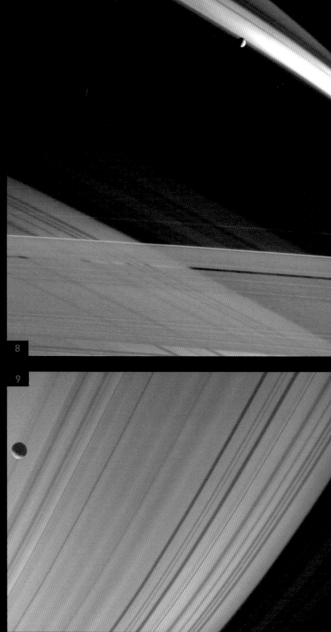

7 ELONGATED SHADOW

Low-angle illumination from the Sun causes the moon Mimas to cast an extremely elongated shadow across Saturn's rings. The shadows were seen as the Sun crossed Saturn's equator during August 2009, bringing a Spring season to the planet's northern hemisphere for the first time in almost 30 years.

8, 9 STUNNING SATURN

Conjunctions of Saturn's rings and its moons, and the slanting of sunlight, provide endlessly variable and fascinating panoramas. Enceladus traverses from dark to light zones in this view from above Saturn's southern hemisphere [8] and Mimas is backdropped by the rings [9] in this view from the south.

1

1 FOUR MOONS

Twice in every Saturnian orbit, the rings are visible edge-on from the point of view of Earth's telescopes. Four moons can be seen, from lower left to upper right: Enceladus, Mimas, Dione and Titan.

2, 3, 4 CHANGEABLE WORLD

The Cassini space probe used infrared cameras to penetrate Titan's thick atmosphere and reveal the moon's surface. The first two images of the same hemisphere [2,3] were taken two and a half months apart. Variations in colour are caused by weather and the changing distribution of various ices on the surface. The final frame [4] is of the opposite hemisphere.

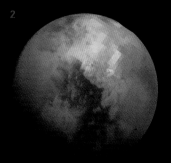

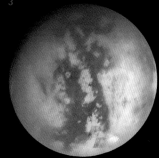

5, 6 THICK ATMOSPHERE

Titan is the only moon in the Solar System to possess a substantial atmosphere [5]. An image taken by Cassini shows its haziness and rich colour, while this edge-on view [6] shows its impressive thickness.

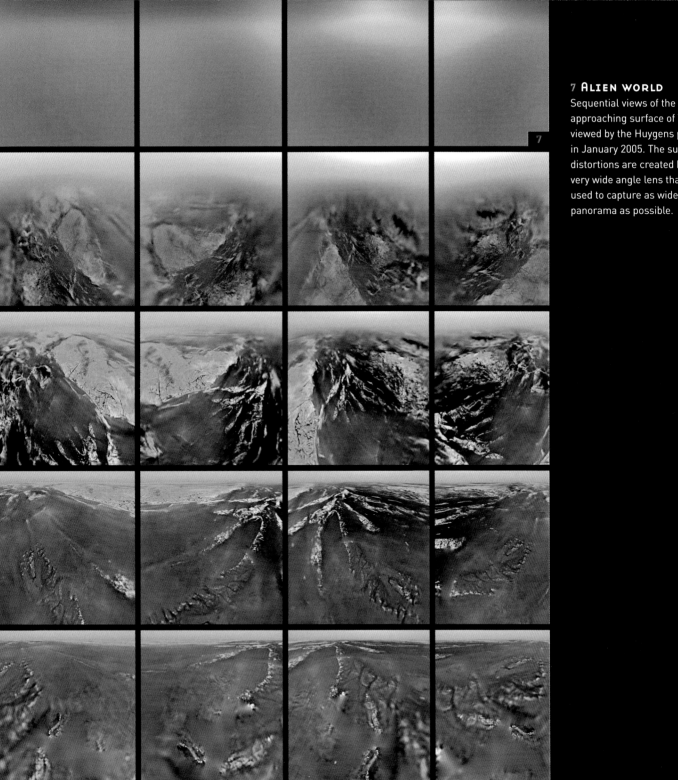

7 ALIEN WORLD
Sequential views of the approaching surface of Titan viewed by the Huygens probe in January 2005. The surreal distortions are created by the very wide angle lens that was used to capture as wide a panorama as possible.

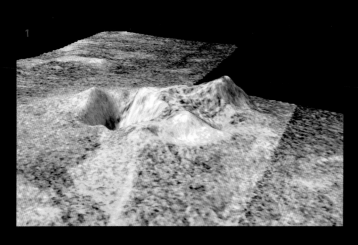

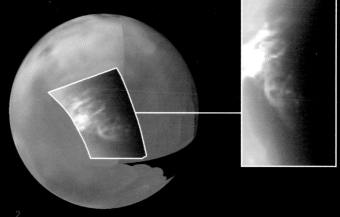

1 Titanic volcano

Radar mapping of an area of Titan known as Sotra Facula provides good evidence for an ice volcano, or 'cryovolcano'. Multiple craters are as deep as 1,500 m (5,000 ft). The image also shows finger-like flows of ejected material. In this colour scheme, blue suggests the presence of exposed ice.

2 Sluggish weather

Infrared light from Titan shows a large cloud formation in the moon's south polar region. The clouds accumulate and drift much like clouds on Earth, but in a slower, more lingering fashion. This is a consequence of the extremely cold conditions.

3 Perspective on Titan

This simulated view, generated by data from the Huygens probe, shows what Titan's surface would look like through human eyes, and without the space probe's wide angle distortion. The liquid lake in the foreground seems as clear as water, but is in fact methane.

4 Huygens probe

Cassini carried a small landing probe, called Huygens, designed to reach Titan's surface. This distorted fish-eye view is what Huygens saw as it fell through Titan's atmosphere and approached the surface. Rough, bright icy terrain surrounds dark, flat lakes of liquid methane.

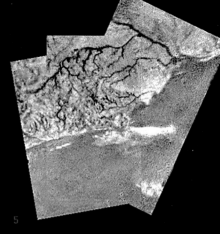

5, 6 TITAN'S LAKES AND RIVERS

Erosion features and channels carved by liquid are evident in this view of Titan from the descending Huygens probe [5]. Subsequent radar mapping shows flat lakes of liquid methane [6].

1 THIRD MAJOR MOON

Tethys is larger than either
Mimas or Enceladus. Its surface
is relatively bright, but heavily
cratered. Obviously the crust
does not rapidly reshape itself
like the icy surface of Enceladus.

2 TETHYS TRENCH

Ithaca Chasma is a long trench
or valley extending around much
of the moon's surface, possibly
created when layers of collapsed
under their own weight.
Alternatively, it may have formed
from rock, as the crust cooled
soon after its formation so that
all we see is the approximate
ice-encrusted outline of a much
deeper flaw.

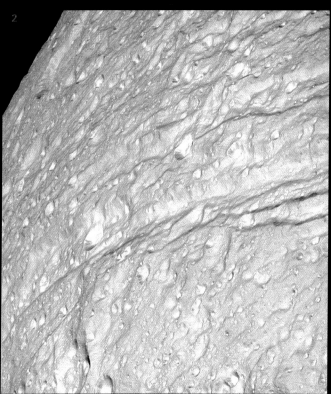

3, 4 THREE AT ONCE

This Cassini image shows Dione,
Tethys and tiny Pandora [3] from
just above the plane of Saturn's
rings. The apparent closeness of
the moons is deceptive, a product
of Cassini's powerful telescopic
system. Dione and Pandora are
in front of the rings, while Tethys
is behind them. A closer view of
Dione [4] shows a network of
bright streaks on the left limb.
These are cliffs of ice.

1, 2, 3 JETS OF ICE

Plumes of ice material ejected from fissures in the crust of Enceladus can reach at least 1,000 km (600 miles) into space [1, 2, 3]. The ice geysers arise from near-surface pockets of liquid water with temperatures substantially higher than the moon's deep-frozen surface.

5, 6 TIGER STRIPES

Because of its bright icy exterior, Enceladus [5] has one of the highest albedos, or levels of reflectivity, of any object in the Solar System. The grooves and ridges in the southern hemisphere [6] seem to be where the crust comes under the greatest stress, probably as a result of Saturn's tidal pull.

4 ICE JET DETAILS

An ice plume ejected from Enceladus sends particles streaming into space hundreds of kilometres above the South Pole of this highly active moon. Some of the particles escape to form the diffuse outermost ring around Saturn. This colour-coded image was processed to enhance faint signals, making the fainter, larger-scale components of the plume easier to see.

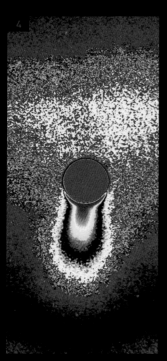

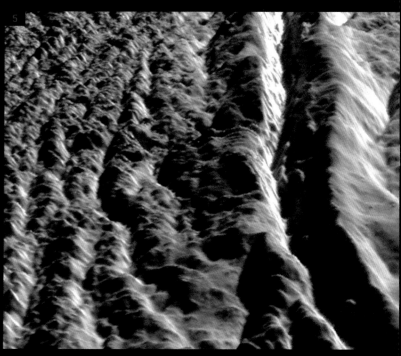

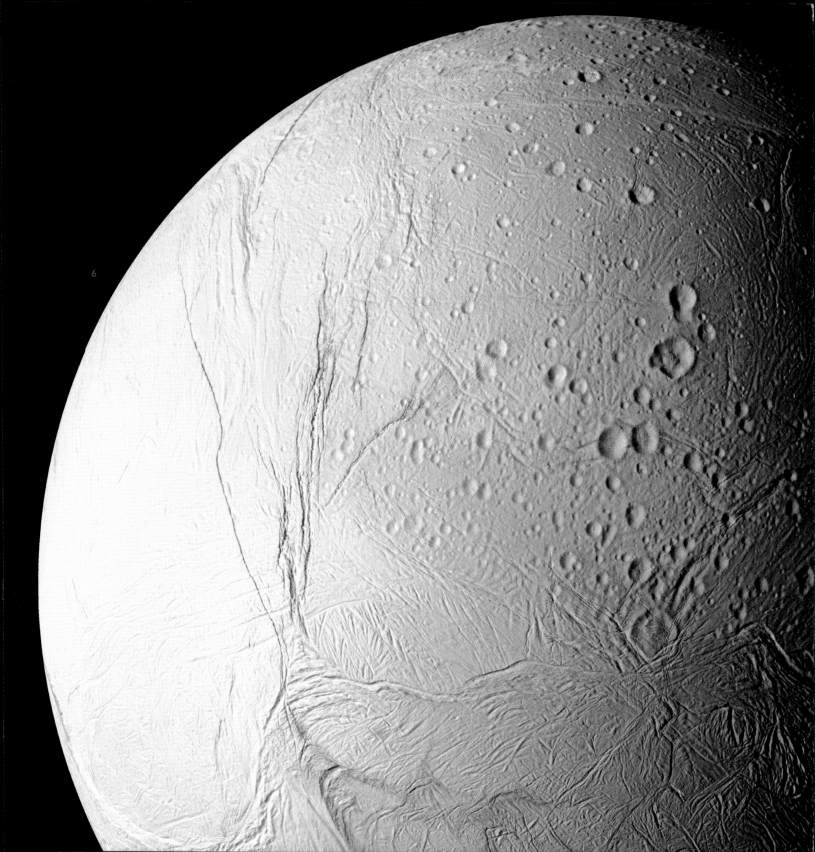

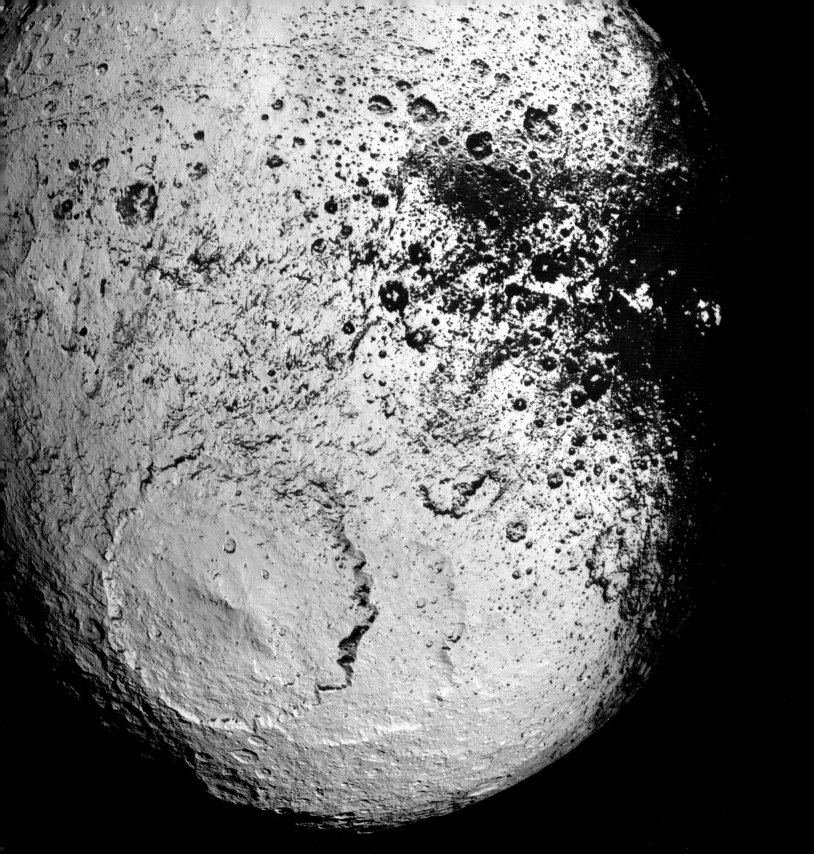

← 1 Light and dark

Iapetus is unusual in that large areas of its surface are markedly dark, and others very light. This moon turns on its axis once every 80 days. The slow turning has complex consequences, and it seems that dust from the rings, and from other moons, deposits onto one hemisphere of Iapetus, but not the other.

2, 3 Iapetus from the north

In this mosaic of two high resolution images taken during Cassini's New Year's Eve 2004 flyby of Iapetus on New Year's Eve 2004, the north pole is approximately 15 degrees below horizontal on the right. The dark deposits become patchy and diffuse as the surface transitions into much brighter and icier terrain near the pole.

5 Landslide

A tremendous collapse of material on Iapetus created this distinct ridge, known as Cassini Regio. The fact that it extends so far across the surface suggests that the material may be very fine-grained. The material appears to have collapsed from a scarp 15 km (9 miles) high.

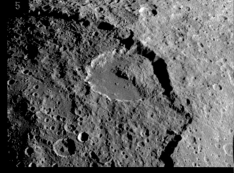

5

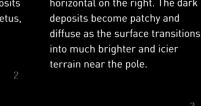

2

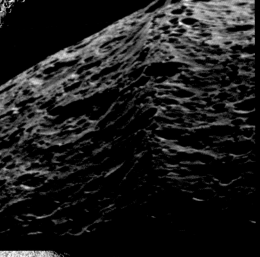

3

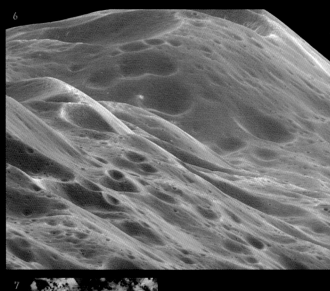

6

7

8

4 Half and half

Iapetus is more intriguing because it has a large and unusually bright patch on its surface, and a strange seam-like ridge along its equator, almost as though the moon's crust is in two halves, joined like the halves of a walnut shell.

6, 7, 8 Iapetus images

High resolution scans (6, 7, 8) of the transition region between the bright and dark hemispheres on Iapetus reveals bright water ice spotted with darker carbon-rich material. The images were taken in September 2007 as the Cassini spacecraft approached to within 5,260 km (3,270 miles) of Iapetus. Another close-up view taken on that same flyby shows mountainous terrain reaching about 10 km (6 miles) high along the unique equatorial ridge of Iapetus.

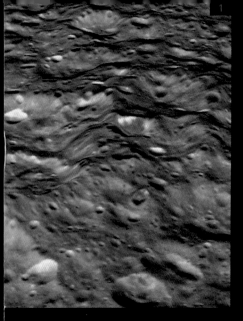

1, 2 SWEEPER MOON

The Saturnian moon Rhea [2] is little more than a ball of ice with a small rocky core. The obvious impact scar on Mimas [1], some 140 km (87 miles) wide, is called Herschel Crater. Mimas may be small, but its gravitational influence is largely responsible for keeping the Cassini Division of Saturn's rings relatively clear.

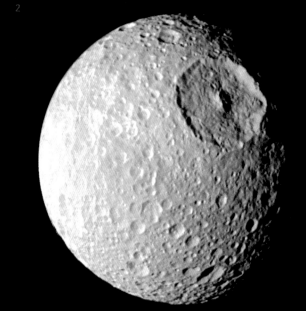

3, 4 THIN AIR

Rhea and the smaller moon Epimetheus are seen partially obscured by Saturn's rings [3] and, in a later orbital space probe pass, Rhea is backdropped more clearly against the rings [4]. Rhea has an extremely thin oxygen atmosphere; almost, but not quite, undetectable. In March 2010 the Cassini probe scooped a tiny sample of oxygen and carbon dioxide while passing overhead at an altitude of 97 km (60 miles). The wispy atmosphere is sustained by high-energy particles slamming into Rhea's surface and kicking up atoms and molecules.

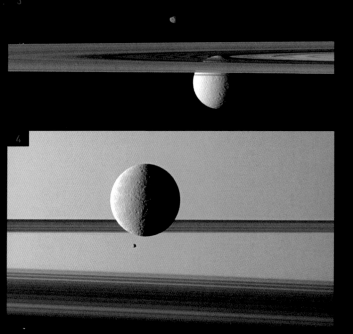

5 HERSCHEL CLOSE UP

Mimas's largest impact crater spans a third of the moon's diameter. If the object responsible for this dent had been any larger, it would have shattered Mimas into fragments, probably creating another Saturnian ring. The walls of Herschel Crater are approximately 5 km (3 miles) high, but parts of the floor descend 6 km (10 miles) below the rim.

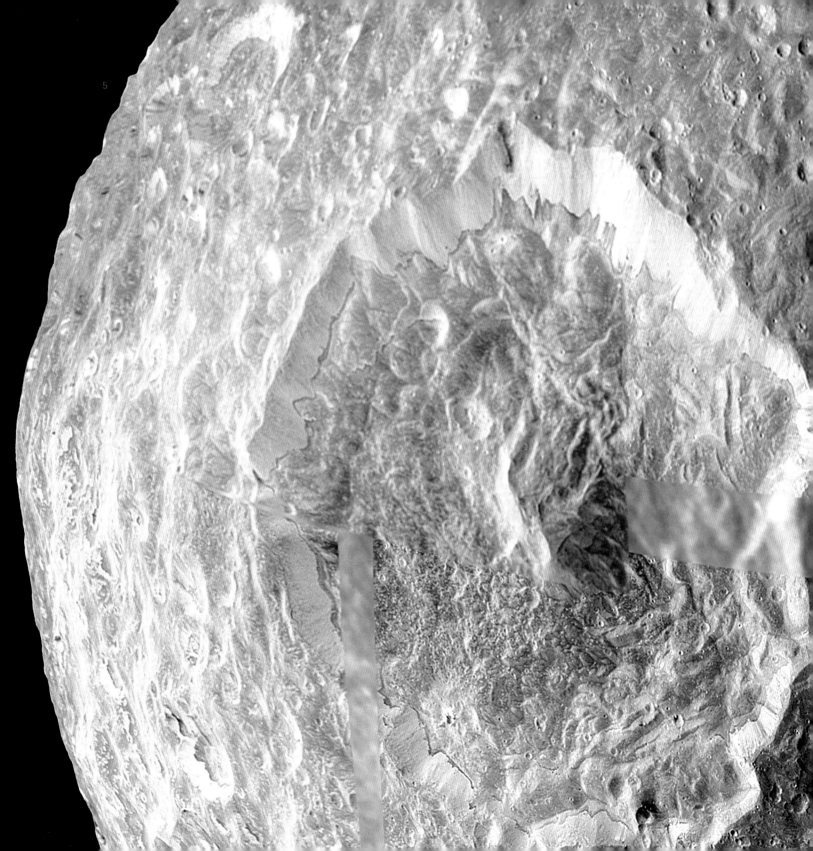

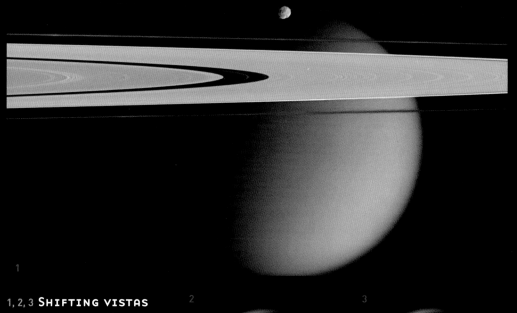

1, 2, 3 SHIFTING VISTAS

One Saturnian moon slips behind another in this time-lapsed pair of images [2,3] taken during 2004. Titan's dense and hazy atmosphere is obvious. As Cassini sweeps through the Saturnian system, its perspective shifts [1] and it sees small, battered Epimetheus and smog-enshrouded Titan, with Saturn's A and F rings in the foreground.

4 CONJUNCTION

The Cassini spacecraft observes a pair of Saturn's moons, with the hazy sphere of comparatively giant Titan behind the smaller Saturnian moon, Tethys.

5 SPONGY MOON

Hyperion is strewn with strange craters and has a very peculiar sponge-like surface. The bottom of most craters contain some type of unknown dark material. Hyperion is small, at about 250 km (155 miles) across. It rotates chaotically, and has a density so low that it may hide a vast system of caverns inside.

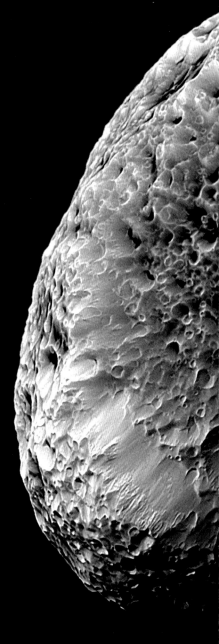

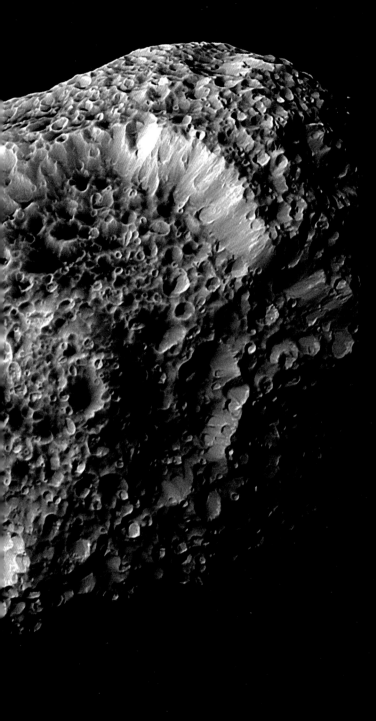

6, 7 MINOR MOONS

Pandora [6] is a fragment from
a larger object that was pulled
apart by Saturn's gravity.
Epimetheus [7] is a so-called
'shepherd moon'. Its gravity,
though very weak, helps to define
the shape of Saturn's faint outer
'F' Ring.

6

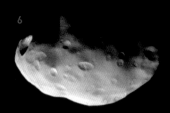

7

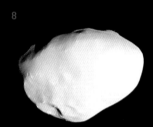

8, 9 TROJAN MOONS

Telesto [8] is known as a Tethys
Trojan because, together with
Calypso [9], it circles Saturn in a
similar orbit as the moon Tethys.
Telesto is about 24 km (15 miles)
across and appears to have a
smooth, icy surface. It does
not show the signs of intense
cratering seen on many of
Saturn's other moons.

8

9

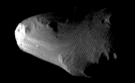

10

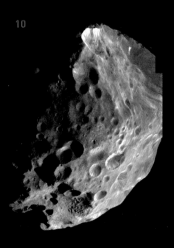

10 CAPTURED ASTEROID

As its appearance suggests,
Phoebe is probably a stray
asteroid caught by Saturn's
gravitational field, but traces
of carbon dioxide ice have been
identified on its surface. This is
not a characteristic of other
asteroids.

1 BLUE-GREEN BALL

At first glance one of the blander-looking planets, Uranus has some of the brightest clouds in the solar system, and its own faint ring system. The atmosphere is mainly hydrogen and helium, with a small amount of methane and traces of water and ammonia. The blue-green colour results from methane absorbing the red portion of the sunlight that strikes the planet, while the blue-green portion is reflected back into space by the cloud tops.

2 TILTED AXIS

Uranus may have been struck by a planet-sized object early in its life and knocked onto its side. Consequently, its axis is tilted to such an extent, its north-south axis almost points towards the Sun. This image shows the similarly tilted faint ring system.

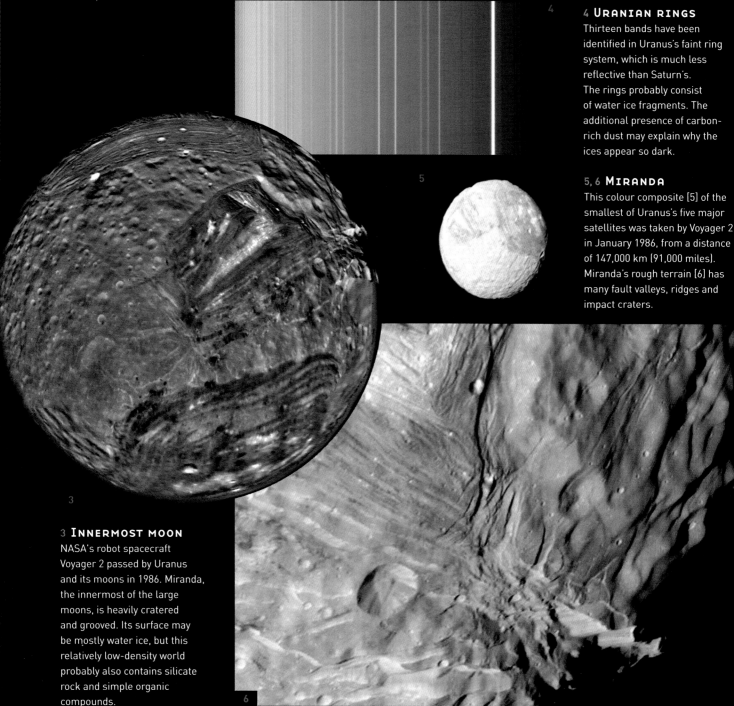

4 URANIAN RINGS

Thirteen bands have been identified in Uranus's faint ring system, which is much less reflective than Saturn's. The rings probably consist of water ice fragments. The additional presence of carbon-rich dust may explain why the ices appear so dark.

5, 6 MIRANDA

This colour composite [5] of the smallest of Uranus's five major satellites was taken by Voyager 2 in January 1986, from a distance of 147,000 km (91,000 miles). Miranda's rough terrain [6] has many fault valleys, ridges and impact craters.

3 INNERMOST MOON

NASA's robot spacecraft Voyager 2 passed by Uranus and its moons in 1986. Miranda, the innermost of the large moons, is heavily cratered and grooved. Its surface may be mostly water ice, but this relatively low-density world probably also contains silicate rock and simple organic compounds.

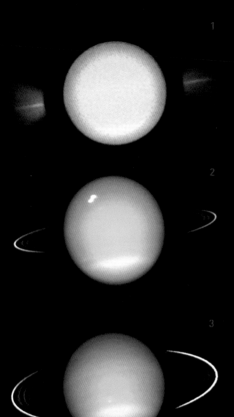

5 NEPTUNE

Neptune, a similar world to Uranus, orbits the Sun once every 165 years. Unlike all other planets in the Solar System, it is invisible to the naked eye because of its extreme distance from the Earth. It was first seen in a telescope in 1846. Astronomers at the time knew it must exist, or else the equations describing the orbits of the other planets would have been wrong. Neptune's orbit marks the boundary where the Solar System's principal realm of major planets and moons ends and a vast outer region sparsely occupied by 'dwarf' planets and cometary wanderers begins.

1, 2, 3 BRIGHT CLOUDS

Infrared images acquired at intervals by the Hubble Space Telescope show movements of clouds on Uranus, and variations in their distribution. Notice the especially bright cloud in the middle image.

4 MAJOR FIVE

A montage of Uranus's five largest satellites. From left to right in order of increasing distance from Uranus are Miranda, Ariel, Umbriel, Titania and Oberon. The two largest, Oberon and Titania, are about half the size of Earth's Moon, or roughly 1,600 km (1,000 miles) in diameter. Data is incomplete for Miranda and Ariel, and the grey circles depict the missing areas.

6 DARK SPOTS

Clouds on Neptune move at different speeds, depending on their altitude. The Great Dark Spot at the centre of this image circled the planet more slowly than the Small Dark Spot at bottom. Almost everything we know about Neptune is derived from the Voyager 2 flyby in 1989. More recent Hubble Telescope observations suggest that the Great and Small Dark Spots were transient.

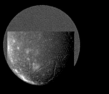
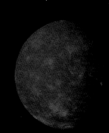
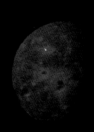

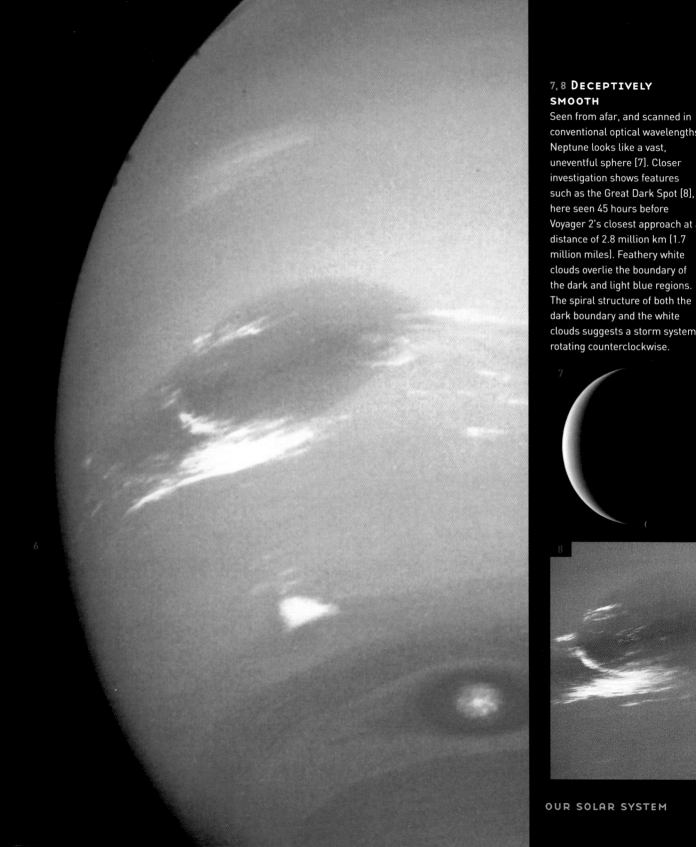

7, 8 DECEPTIVELY SMOOTH

Seen from afar, and scanned in conventional optical wavelengths, Neptune looks like a vast, uneventful sphere [7]. Closer investigation shows features such as the Great Dark Spot [8], here seen 45 hours before Voyager 2's closest approach at a distance of 2.8 million km (1.7 million miles). Feathery white clouds overlie the boundary of the dark and light blue regions. The spiral structure of both the dark boundary and the white clouds suggests a storm system rotating counterclockwise.

7

8

6

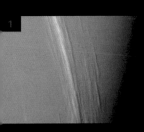

1 Streamers on Neptune
High-altitude clouds often stretch into long streamers shaped by strong winds and the Coriolis forces of Neptune's rotation. The streamers cast shadows on the blue layers of atmosphere beneath them.

2 Deceptive smudges
At first glance, the dark streaks and fuzzy spots in this view of Triton (Neptune's largest moon) look like surface colouration. They are actually the shadows cast by plumes of nitrogen gas and dust ascending into Triton's thin atmosphere.

3 Natural gas
The colouring of both Neptune and Uranus is affected by trace quantities of atmospheric methane gas. The Voyager 2 space probe identified high-altitude methane on Uranus.

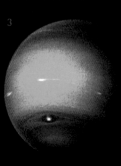

4, 5 PINK MOON

Features as small as 100 km (62 miles) across can be seen in this colour image of Triton [5], photographed by Voyager 2 in August 1989, while it was still 5.4 million km (3.3 million miles) from Neptune. Triton's pinkish colour may be due to chemical products created by solar radiation working on methane gas and ice. As Voyager approaches, the bright southern hemisphere of Triton, which fills most of this frame [4] is clearly reddish-pink in tone.

6 ROUGH LANDSCAPE

The surface of Triton is very rugged, scarred by chunks of ice, crustal faults, volcanic pits and lava flows composed of water and other ices. The surface is also extremely young and sparsely cratered, and is still geologically active today. This scene is 500 km (310 miles) across.

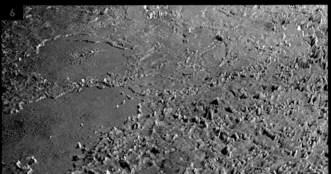

7 FAINT WINDS

In this composite of Voyager 2 images taken through violet, green and ultraviolet filters, a layer of pinkish material stretches across the centre of the image. The elongated dark streaks may represent dusty materials blown in the same direction by prevailing winds, and the white material may be frost deposits. Triton's extremely tenuous nitrogen atmosphere extends 800 km (500 miles) above the surface.

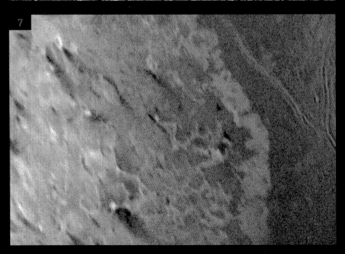

1, 2 Ice dwarf

Discovered in 2003 in an orbit far more distant from the Sun than Pluto's, Eris [1] caused a stir among astronomers, throwing into question what we mean by the word 'planet'. When it became apparent that Eris is larger than Pluto, the International Astronomical Union reclassified Pluto as a dwarf planet. Another observation by Hubble [2] resolves Pluto and Charon in greater detail. The image was taken by the European Space Agency's Faint Object Camera in February 1994 when Pluto was 4.4 billion km (2.6 billion miles) from Earth, or nearly 30 times the separation between Earth and the Sun.

3 Kuiper belt object

Quaoar orbits the Sun at a distance of approximately 6 billion km (10 billion miles), in a region known as the Kuiper Belt. With a diameter of 1,300 km (800 miles) Quaoar is significantly larger than Ceres, the largest asteroid.

7 Farthest outpost →

Sedna, the most distant object so far identified in the Solar System, takes 10,500 years to orbit the Sun. It currently lies about three times as far from the Sun as Neptune, but its orbit will eventually take it thirty times further away. Almost nothing is known about Sedna, except that its surface is almost as red as Mars.

4, 5, 6 Rare details

Intensive computer analysis of extremely faint and tiny images has revealed approximate details on the surface of Pluto. This scientific challenge is the equivalent of reading the manufacturer's name printed on a golf ball 50 km (30 miles) away. Hubble observed almost the entire surface of Pluto as it rotated through its 6.4-day period. Pluto appears to have greater large-scale contrasts than any planet except Earth. Most of the features are probably produced by frosts migrating across the surface, and chemical byproducts deposited out of Pluto's nitrogen-methane atmosphere.

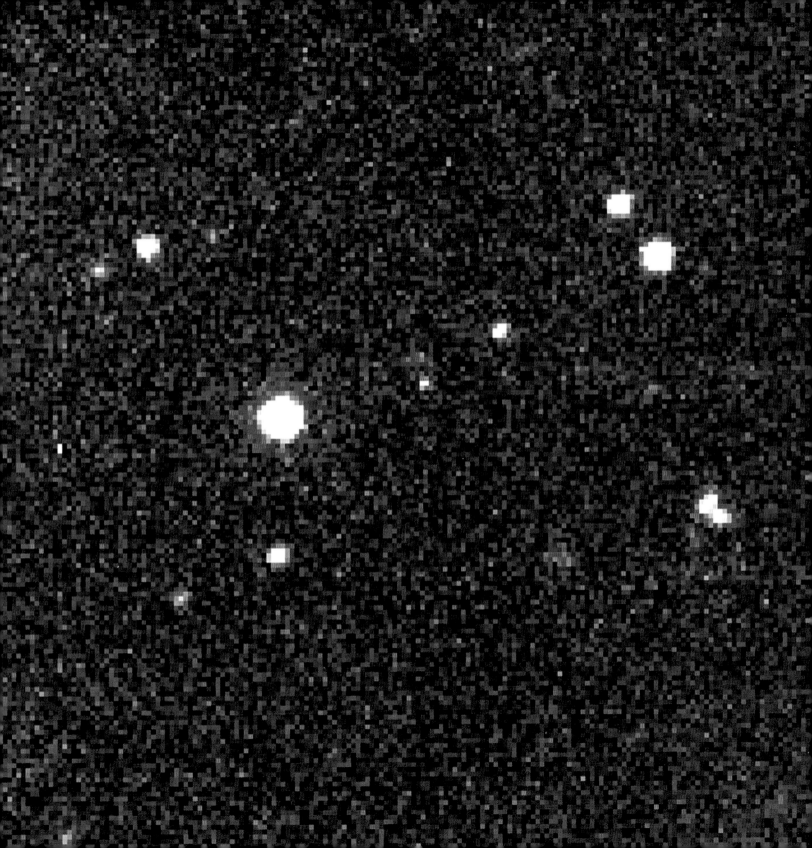

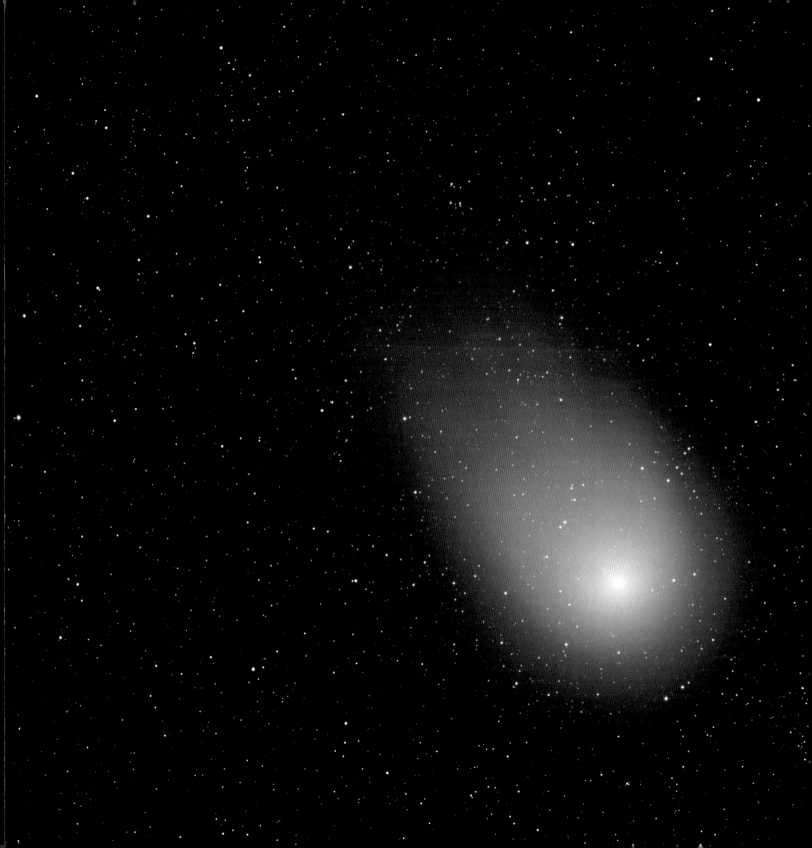

1 NEAT

This comet was discovered by NASA's Near-Earth Asteroid Tracking (NEAT) project in 2001, three years before it passed between the Earth and the Sun, at which point it was easily observed.

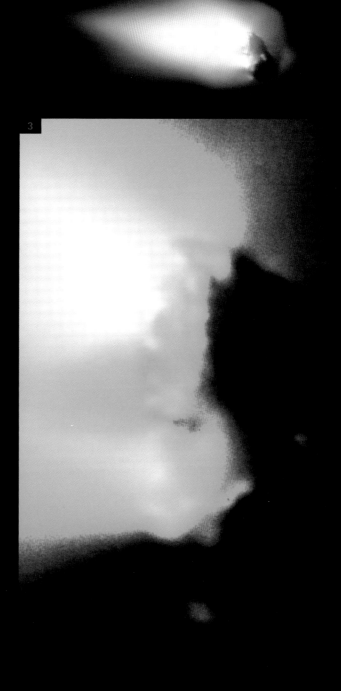

2, 3 FAMOUS COMET

Everyone has heard of Halley's Comet [2]. It passes through the inner Solar System once every 76 years, and was first recorded in the year 240 BC. This close view of its nucleus was taken by the Giotto probe in March 1986. Proximity to the Sun causes volatile compounds (substances with low boiling points, such as water, carbon monoxide, carbon dioxide and other ices) to evaporate from the surface, creating the characteristic glowing tail, which always points away from the Sun. Another view [3] shows dark surface features on the nucleus, with gas and dust flowing into the tail, or 'coma', at left. The debris left behind by Halley is responsible for the Orionids meteor shower in October of every year, and the Eta Aquarids meteor shower every May.

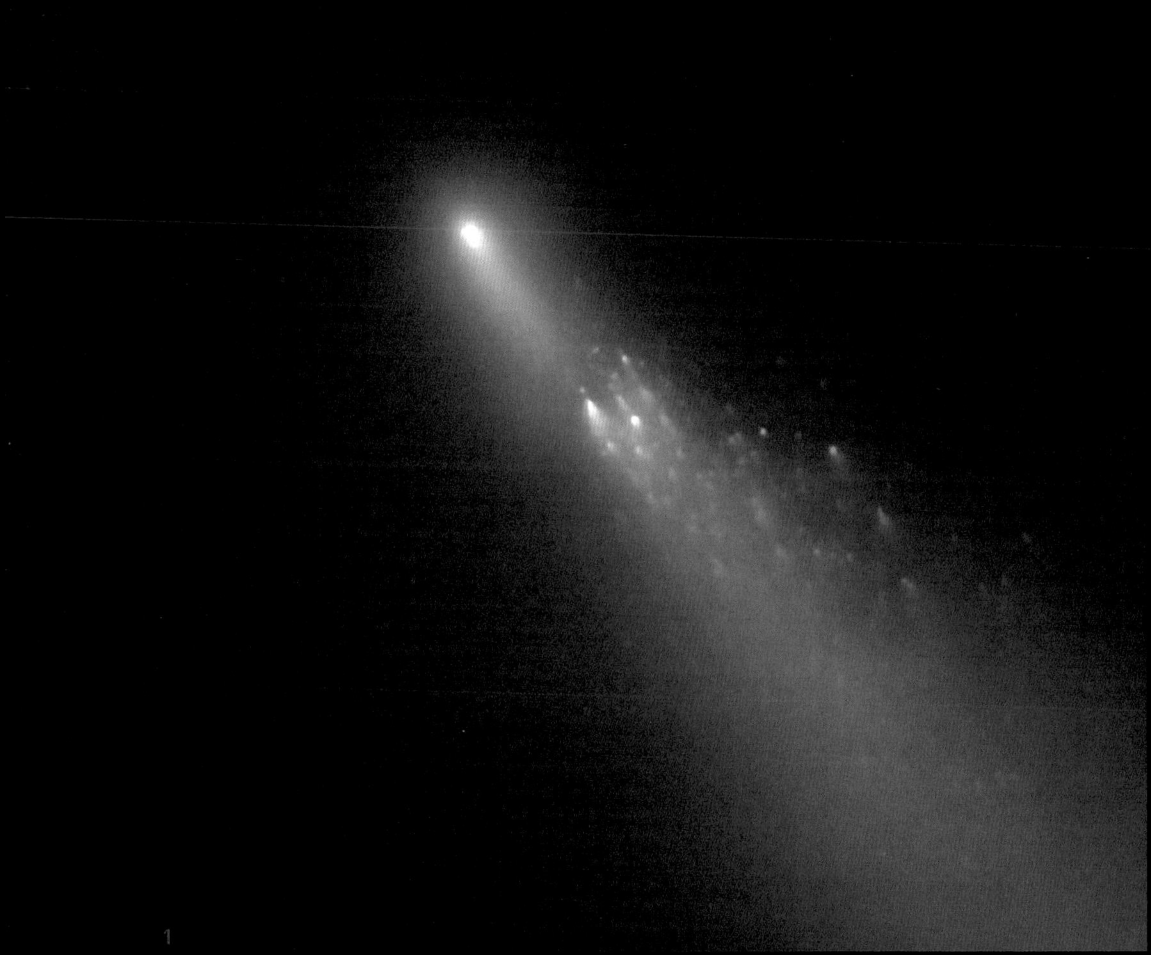

1 BROKEN COMET

Discovered in 1930, comet
Schwassmann-Wachmann 3
orbits the Sun once every 5.4
years, coming to within 776
million km (482 million miles)
at closest approach. It began to
break up in 1995, splitting into
four fragments. By 2006 it had
disintegrated into more than a
dozen pieces. It may not survive
its next swing past the Sun.

2 SCHWASSMANN-WACHMANN

This infrared image from NASA's Spitzer Space Telescope shows Schwassman-Wachmann 3 skimming along a trail of its own debris. The flame-like objects are the comet's fragments and their tails, while the dusty debris trail is the line bridging the fragments. Debris ranges in size from pebbles to large boulders.

3

3 GREEN COMET

Photographed in 2007, Comet Lulin's green colour comes from the gases that make up its Jupiter-sized atmosphere. Jets spewing from nucleus contain cyanide a poisonous chemical found in many comets.

1 SPECTACULAR TAIL
In late 2006, Comet McNaught became visible to the naked eye. It was the brightest comet seen in the last 40 years, but it is currently heading out of the Solar System, and will never return.

2 WILD COMET
The Stardust probe flew past comet Wild 2 in January 2004. A brief exposure of surface detail was overlaid with a long exposure of the much fainter jets. Together, the images show an intensely active comet leaving a trail millions of kilometres long.

3 BRIGHT PERFORMER
Discovered in 1862, comet Holmes made a good display in 2007, when a rapid outburst transformed it from a faint comet quietly orbiting the Sun with a period of about seven years, to a naked-eye comet rivalling the brighter stars.

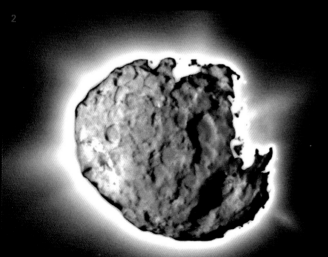

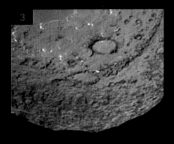

1, 2, 3, 4 DEEP IMPACT

Approaching comet Tempel 1 at 10 km (6 miles) per second, the Deep Impact probe observes a primordial chunk of Solar System material. Tempel 1's ingredients [1,2] include tiny grains of silicates, iron compounds, hydrocarbons, and clays that must have formed in association with water. The view from Deep Impact's probe [3], 90 seconds before it deliberately crashed into Tempel 1. The bright flash [4] signifies the sudden release of energy triggered by Deep Impact's high-speed collision with Tempel 1.

5, 6 CLOSE APPROACHES

The NASA probe Deep Space 1 approached Comet Borrelly [5] in December 2001. The comet's nucleus is about 8 km (5 miles) long, and has smooth rolling plains, and materials of vastly different reflectance. Images taken at different wavelengths [6] shows dust jets escaping Borrelly's nucleus.

7 TRACKING PROGRESS

Deep Impact's progress is tracked by the powerful Mt. Palomar 508-cm (200 inch) telescope in California, as the probe heads towards Tempel 1.

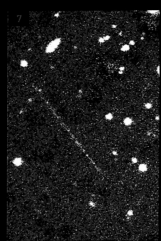

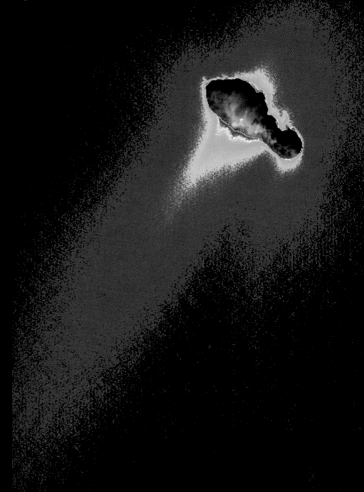

9

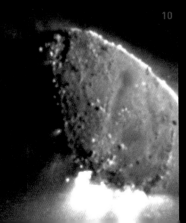

10

11

9, 10, 11 VISIBLE JETS

As seen through Earth-based telescopes in 2010, comet Hartley 2 is little more than a hazy speck of light [9]. Closer inspection by the EPOXI mission spacecraft in November of that year showed jets of dust and gas, indicating its active surface [10]. Sunlight illuminates the nucleus from the right. Hartley 2's jets in even greater detail [11]. What appear to be loose boulders are visible on the nucleus.

1   2

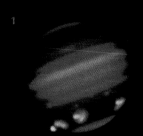

1, 2, 3, 4 CAPTURING AN IMPACT

Comet Shoemaker-Levy 9 was first detected in March 1993. Astronomers watched as it was gradually torn to pieces [3, 4] by Jupiter's strong gravity. In July 1994 a series of fragments slammed into Jupiter, leaving scars in the planet's turbulent atmosphere [1, 2] that took several weeks to dissipate. At least 21 impacts occurred on the 'far side' of Jupiter, but as the planet turned that face towards the Earth, the impact zones were monitored in infra-red, highlighting the immense bursts of energy of the collisions.

4

3

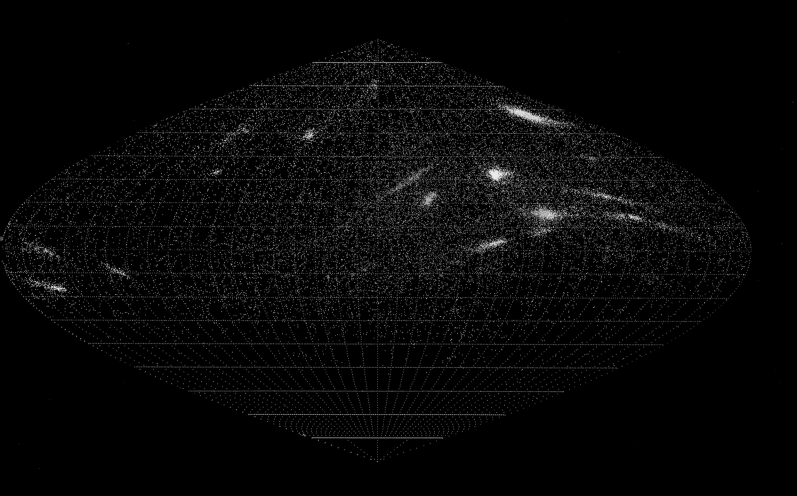

5 Seasonal showers

Visible meteors are typically sand-sized grains of ice and rock fragmented from comets. Distributions of flashes in the night sky are mapped, showing particular concentrations associated with known comet tail debris.

6, 7 Spiral comet

These NASA Hubble Space Telescope pictures of comet Hale-Bopp [6,7] show a remarkable 'pinwheel' pattern and a chunk of free-flying debris near the nucleus. This may be a piece of crust that was ejected into space by the comet's rotation, and which then disintegrated into a bright cloud of particles.

1 Rapid passage

Comet Hyakutake, discovered in January 1996, passed very close to Earth in March of that year. It was moving so rapidly across the night sky that its movement could be detected against the stars in just a few minutes.

2 Encke

NASA's Spitzer Space Telescope shows the comet Encke riding along its pebbly trail of debris between the orbits of Mars and Jupiter. Twin jets of material can also be seen shooting away from the comet in a fan-shaped emission. Encke orbits the Sun every 3.3 years. Every October, Earth passes through its wake, resulting in the well-known Taurid meteor shower.

INTERSTELL

AR SPACE

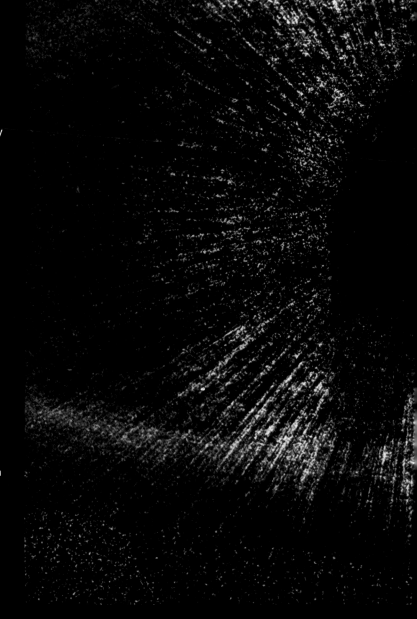

O ur solar system drifts at one end of a spiral arm in a fairly typical galaxy within a universe so vast, there is no easy way for us to grasp the distances involved. The Earth orbits the Sun at a distance of 150 million km (93 million miles). This is a convenient yardstick for comparing distances in and around the Solar System.

The Earth orbits the Sun at a distance of 150 million km (93 million miles). This is a convenient yardstick for comparing distances in and around the Solar System. The Earth-Sun distance is known as an Astronomical Unit (AU). To get an impression of what this really means, imagine the Sun as a bright yellow basketball and the Earth as a small pebble about 30 meters (100 feet) away from it. Pluto's average distance from the Sun is about 40 AUs.

The outermost realms of the solar system, where lonely comets lurk, is somewhere around about 100,000 AUs in diameter. Beyond that already vastly distant realm, even the AU loses its descriptive power as a measure of distance. We have to turn, instead, to the fastest-known entity in the universe. Light travels one AU in eight minutes. When measuring distances in the broader universe, we need to think in terms of how far light can travel in thousands, millions and even billions of years, rather than minutes.

The speed of light is 299,792 km (186,282 miles) per second. In one year, it travels 9,460,800,000,000 km (5,865,696,000,000 miles). Numbers of this size become hard to deal with when written down as simple numerals. For convenience, astronomers call this special distance a 'light year'. The next nearest star to us is Proxima Centauri, 4.2 light years distant. Imagine brilliant, glowing oranges separated from each other by the same distance as New York is from Chicago, and this gives a rough idea of the mind-numbing distances between stars that, in cosmic terms, are next door to each other.

Our galaxy, the Milky Way, contains approximately two hundred billion stars. From one side of the galaxy to the other, the distance is around 100,000 light-years. The light that we capture in our telescopes from stars at the galaxy's farthest perimeter was first emitted while Neanderthal humans still roamed the Earth.

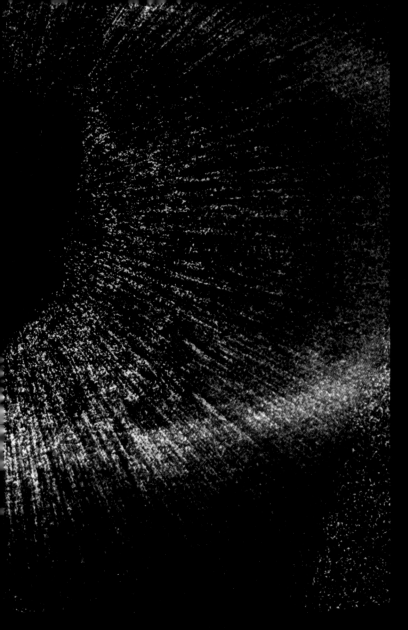

cores in a nuclear fusion process, liberating energy as they do so. Eventually the hydrogen runs out and the core can no longer generate enough outwards radiation to balance the inward pull of gravity. The centre of the star shrinks while the outer layers expand. This makes the star cooler, and it becomes a red giant. Eventually the weakened star collapses and becomes a white dwarf. It may even be that white dwarfs eventually dim yet further, to become black dwarfs, but that final fade-out would be so slow that we may never be able to prove that it happens. At the other end of the extreme, stars that start life very much smaller than the Sun may just dim over time, or perhaps never even ignite in the first place, becoming 'brown dwarfs,' emitting so little light they are almost invisible to our telescopes.

The names and classifications of celestial objects can be confusing, but two systems in particular are very useful for any astronomer, whether amateur or professional. The capital letter 'M' followed by a number, such as in M84 or M87, is a legacy from the French astronomer Charles Messier, a comet hunter who, in 1771, published a list of the positions of about a hundred fuzzy, cloud-like objects in the sky. He knew they could not be comets, because these clouds, or 'nebulae', were fixed against the background of stars. Messier notation is useful for astronomers to identify familiar targets, including vast star-forming clouds of gas and dust, or the smaller but no less impressive shells of gas hurled out by dying stars.

In the 1880s, astronomers working with more powerful telescopes than in Messier's day began The New General Catalogue of Nebulae and Clusters of Stars, known today as the NGC system. As modern instruments peer deeper into the cosmos, resolving exotic phenomena at far greater distances than before, many new astronomical classifications have been invented, often named in honour of the astronomers who first identified typical examples. But the Messier and NGC systems are still invaluable for that most basic of problems: where to point your telescope if you want to observe a particular object.

Stars within the galaxy are born out of the same processes but live and die in different ways. Astronomers use a standard reference system called the Hertzsprung-Russell diagram. Many stars, including our Sun, fall into a category known as the Main Sequence, but stars come in a tremendous range of sizes. The mass of a star in its prime of life determines what happens to it as its hydrogen fuel reserves fade away and it begins, fitfully and violently, to die.

Stars the size of the Sun turn hydrogen into helium in their

THE NEAREST STAR

A certain unimpressive star is so dim, it can only be seen in a telescope. Its main claim to fame is that it is closer than any other star apart from the Sun.

The nearest stellar neighbours to the Sun are three stars that make up a multiple system. To the naked eye the system appears as a single bright star, Alpha Centauri, but this is, in fact, a binary star system, with two stars orbiting around a mutual centre of gravity. Near these is the third member of the system, a faint star known as Proxima Centauri, slightly nearer to the Sun than the other stars in this triple star system.

Proxima Centauri is just over four light years way from us. This little red dwarf star, 18,000 times fainter than the Sun, is so faint that it was only discovered in 1915 and is only visible through a telescope. The star's light output varies randomly over very short periods, so it may be subject to flares or other eruptions.

The 'Proxima' part of the star's name comes from a Latin word meaning 'near'. This unimpressive orb is an unlikely host for a solar system, let alone any planets suitable for life. If any worlds do orbit Proxima Centauri, they are probably dark and cold.

THE NEXT NEAREST STAR *Light from the Sun takes eight minutes to reach Earth. It takes more than four years to reach us from the next nearest star, Proxima Centauri, a red dwarf that flares up dramatically, tripling its brightness in just a matter of minutes.*

1, 2 SIRIUS SYSTEM

Sirius is the brightest star in the night sky, a closely watched celestial beacon throughout recorded history [1]. Modern astronomers know it as a binary star system. The small companion star, Sirius B, is much dimmer and appears so close to the brilliant Sirius A that it was not actually identified until 1862. For orbiting x-ray telescopes [2], the Sirius situation is exactly reversed. The smaller but hotter Sirius B appears as overwhelmingly more intense than its larger partner.

8 light years

3 NEAREST BROWN DWARF

Epsilon Indi B belongs to the 'brown dwarf' category that blurs the divide between very cool, weak stars and giant planets similar to Jupiter. This exceptionally cool, small and dim star is a companion to a well-known brighter star in the southern sky, Epsilon Indi A.

12 light years

1 Beta pictoris planetary ring

This Hubble image, obtained in 2003, shows a dust disc surrounding the star Beta Pictoris. The disk extends at least 38 billion km (24 billion miles) from the star. A secondary disk is circumstantial evidence for the existence of a planet.

64 light years

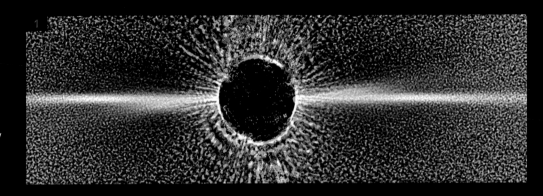

2 HR 8799 star and planets

This incredible picture shows multiple planets orbiting the Sun-like star HR 8799. Each planet is probably several times the mass of Jupiter, but this is proof that complex planetary systems exist: systems that could conceivably contain an Earth-like planet.

130 light years

3 Nearest neutron star

The three smudges left of centre represent the progress of RXJ 185635-3754 across the sky, over three years. This is a fast-moving stellar corpse as small as Manhattan Island but 55 trillion times denser. This is a neutron star, the wreckage from a supernova explosion, and it is comparatively close to us at a mere 180 light years away.

180 light years

4 Exoplanet storm

These computer-generated images chart the development of severe weather patterns on exoplanet HD 80606b. An exoplanet is a planet that orbits a star other than our Sun. In 2007 the Spitzer Space Telescope monitored the heat radiating from the planet over a period of five days. The reflected starlight appears blue because the planet is a very efficient absorber of red light.

190 light years

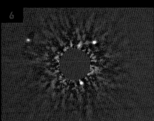

5, 6 PLANET-HUNTING

In 2004 European astronomers captured the first visual image of a planet orbiting another star. The picture shows a dim, red point of light [5]. This is a young, giant planet on a similar scale to Jupiter. It orbits a 'failed' brown dwarf star. In 2008 astronomers made an image of a more Sun-like star, HR 8799, 130 light years away. It clearly has multiple planets [6].

200 light years

7 IRA'S GHOST

Nebula IRAS 05437+2502, otherwise known as Ira's Ghost, is a small star-forming region filled with dark dust. The glowing arc may have been created by a massive star that somehow attained a high velocity and has now left the nebula. Ira's Ghost spans only one twentieth of a full Moon toward the constellation of the Bull (Taurus).

380 light years

← 1, 2, 3, 4 SEVEN SISTERS

Hurtling through a cosmic dust cloud some 400 light-years away, the beautiful and familiar Pleiades or Seven Sisters star cluster [1] is famous for its striking blue nebulae. The clouds are revealed in yet more detail in infrared [2]. The densest regions are highlighted in yellow and red, and the thinnest in green. In the dusty sky toward the constellation Taurus and the Orion Arm of our Milky Way Galaxy, the Pleiades are at the upper left [3] while a closer inspection [4] picks out the individual stars amid their shrouds of dust.

400 light years

STAR
FORMING
REGIONS

The Hubble Space Telescope and other powerful instruments confirm that star formation is not just something that happened in the distant past. It is still happening throughout the galaxy.

All stars are formed by accretion processes very similar to those that created our Sun. Thin wisps of gas and dust are pulled together by gravitation, and as the mass and gravity of the accreted lumps becomes ever greater, so more material is pulled inwards, and the accretion, or protostar, assumes a spherical shape. Eventually the gravitation and the internal pressures becomes so great, nuclear fusion begins, and the protostar flares into life as a fully fledged star.

We also see planetary accretion discs forming around new suns very like our own. Just as often, we see monstrous stars too powerful to support planets. Their accretion discs simply get blasted away by the excessive radiation. What we do know is that destruction and creation walk hand in hand throughout the Milky Way. We know that the formation of solar systems and planetary worlds much like our own is not merely possible, but commonplace. We can speculate that some of them might be Earth-like, and host to living creatures, but proof of this awaits some future discovery.

WITCH HEAD NEBULA *Otherwise catalogued as IC2118, this suggestively shaped reflection nebula is associated with the bright star Rigel in the constellation Orion. Fine dust in the nebula reflects Rigel's light. The dust grains reflect blue light more efficiently than red (664 light years).*

1, 2, RHO OPHIUCHI CLOUD

Dust clouds and embedded newborn stars glow at infrared wavelengths in this false-colour view from the Spitzer Space Telescope [1] and in broader wavelengths from the Hubble [2]. The view spans about 5 light years. After forming along a large cloud of cold molecular hydrogen gas, newborn stars heat the surrounding dust to produce the infrared glow. An exploration of the region in penetrating infrared light has detected some 300 emerging and newly formed stars.

400 light years

1, 2, 3 **MIRA'S TAIL**

NASA's Galaxy Evolution Explorer satellite discovered an exceptionally long comet-like tail of material trailing behind Mira A. The tail is 13 light years long, or about 20,000 times the average distance of Pluto from the Sun. Nothing like this has ever been seen before around a star. The tail is only visible in ultraviolet light [1] and does not show up in visible light [2]. Mira A is losing gas rapidly from its upper atmosphere, and Mira B exerts a gravitational pull that creates a gaseous bridge between the two stars. The gas accumulates in an accretion disk around Mira B [3]. Collisions between rapidly moving particles in the disk produce x-rays [insets].

420 light years

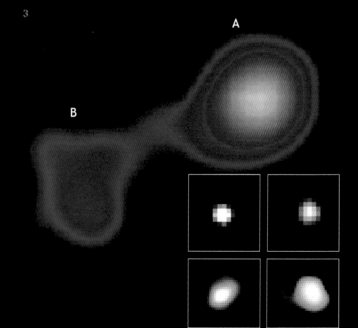

5 **CORONA AUSTRALIS**

A characteristic blue colour is produced as light from hot stars is reflected by cosmic dust. The smaller yellowish nebula NGC 6729 surrounds young variable star R Coronae Australis. Magnificent globular star cluster NGC 6723 is at the upper right corner of the view. It actually lies nearly 30,000 light years away, far behind Corona Australis.

420 light years

1 Planet orbiting star

Located toward the constellation Scorpius, the parent star, catalogued as 1RXS J160929.1-210524, is slightly less massive and a little cooler than the Sun, so it is especially exciting to note the presence of at least one associated planet, even if it is ten times larger than Jupiter.

500 light years

2, 3, 4 XZ Tauri binary plasma burst

This is a system of two stars orbiting each other, and separated by about 6 billion km (4 billion miles), about the distance between the Sun and Pluto. Hubble watched over five years as a bubble of hot, glowing gas extended nearly 96 billion km (60 billion miles) from the system [2,3,4]. Hubble did not find which of the two stars was responsible.

500 light years

5 Chameleon Complex

Thin wispy shells of carbon dust from dead stars are illuminated by two bright young stars, scattering blue light to form this reflection nebula. Surrounding molecular clouds containing thicker dust block the light from more distant stars.

500 light years

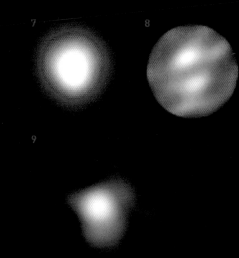

6, 7, 8, 9 BETELGEUSE
This red supergiant star in the
constellation Orion [6] is
shrinking fast. Betelgeuse is ten
times the diameter of the Earth's
orbit around the Sun. In just
fifteen years, its diameter has
shortened by a distance equal to
the orbit of Venus. Enhanced
analysis of its light [7,8] resolves
some surface variations,
including two large bright spots,
enormous convective cells rising
from below the supergiant's
surface. The sharpest image ever
of Betelgeuse [9] was obtained by
the Very Large Telescope in
Chile, and clearly shows its
escaping plumes of gas.
640 light years

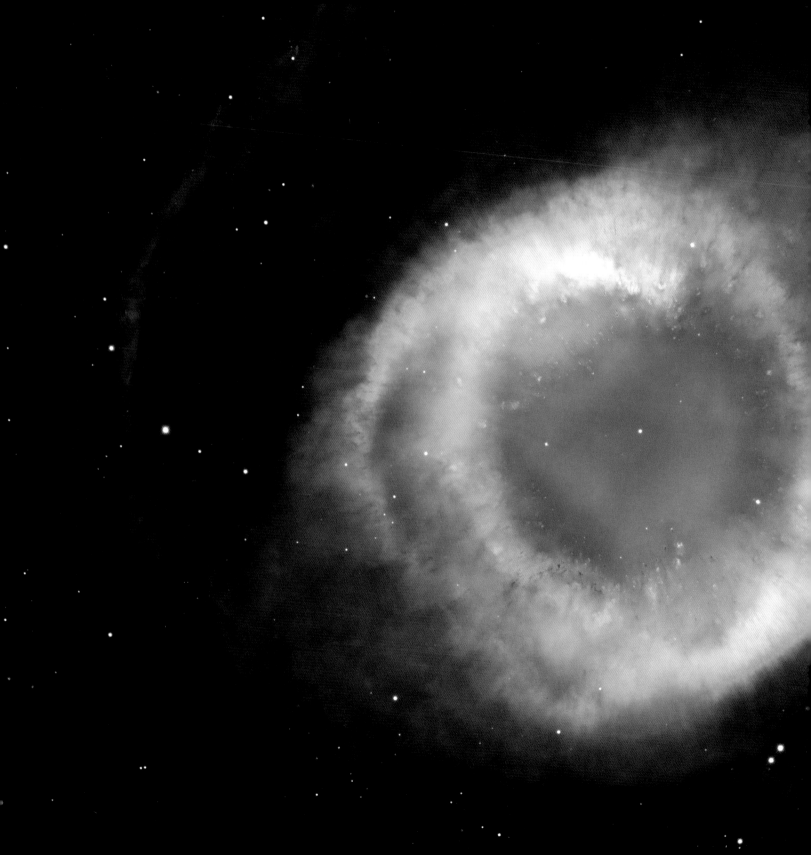

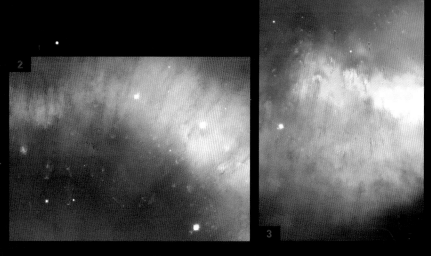

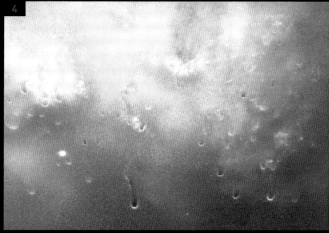

2

3

4

1

1, 2, 3, 4 HELIX NEBULA

This barrel of glowing gas [1] is
being expelled from a star in the
final phases of its life. Its core is
destined to become a white dwarf
star. Its light is so energetic it
causes the surrounding expelled
gas to fluoresce. In these details
[2,3], blue reveals the presence
of oxygen, while red and yellow
show nitrogen and hydrogen.
The edges of the nebula are
populated by comet-like knots of
dense material [4]. Each 'head' is
roughly twice the size of the solar
system.

600 light years

1, 2 NGC 1514
DYING STAR

This puffy, dying star, or
'planetary nebula', is actually
a pair of stars. One is a dying
giant heavier and hotter than our
Sun, and the other was once an
even larger star that has now
contracted into a white dwarf.
This view [1] is from a visible-
light telescope, while the
accompanying image [2] is
infrared light, as seen by
NASA's Wide-field Infrared
Survey Explorer.

600 light years

3, 4, 5 BHR 71 CLOUD

Two young stars are destroying
the dust cloud that gave birth
to them, with powerful jets of
radiation. The stars are in a
cosmic cloud called BHR 71. In
visible light [3] BHR 71 is just a
large black structure. In infrared
light [4] the baby stars are shown
as bright yellow smudges toward
the centre. The combined
visible-light and infrared
composite [5] shows that a
young star's powerful jet is
responsible for the rupture at
the bottom of the dense cloud.

600 light years

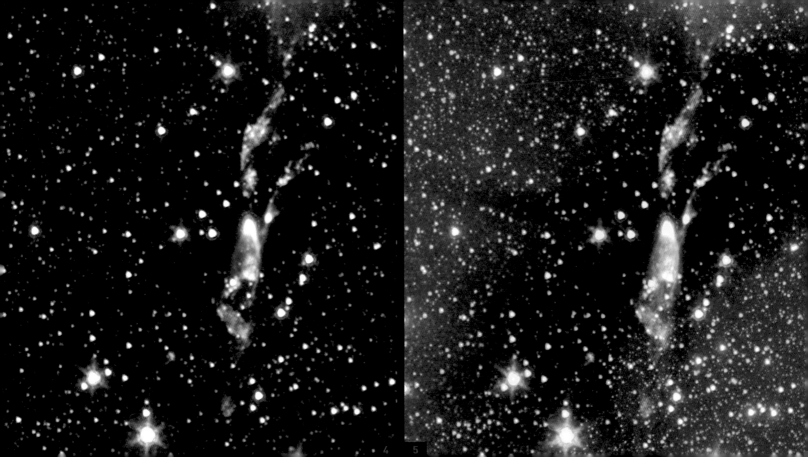

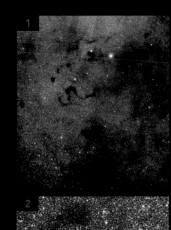

1, 2, 3 Snake nebula

Looking into the densely populated centre of the Milky Way [1] we see a concentration of obscuring dust and dark matter [2, 3] blocking the light from those central stars. The Snake Nebula is also listed as Barnard 72, one of 182 dark regions catalogued in the early 20th century by the American astronomer Edward Barnard.

650 light years

4 Witch Head nebula

Otherwise catalogued as IC2118, this suggestively shaped reflection nebula is associated with the bright star Rigel in the constellation Orion. Fine dust in the nebula reflects Rigel's light. The dust grains reflect blue light more efficiently than red.

444 light years

1, 2 ORION'S BELT

Alnitak, Alnilam and Mintaka, are
the bright bluish stars from east
to west (left to right) along the
diagonal in this view of a familiar
part of the Orion constellation
[1]. These three blue supergiant
stars are hotter and much more
massive than the Sun. A
spectacular wide-field view of
the Orion region [2] shows the
Flame Nebula.

800 light years approx

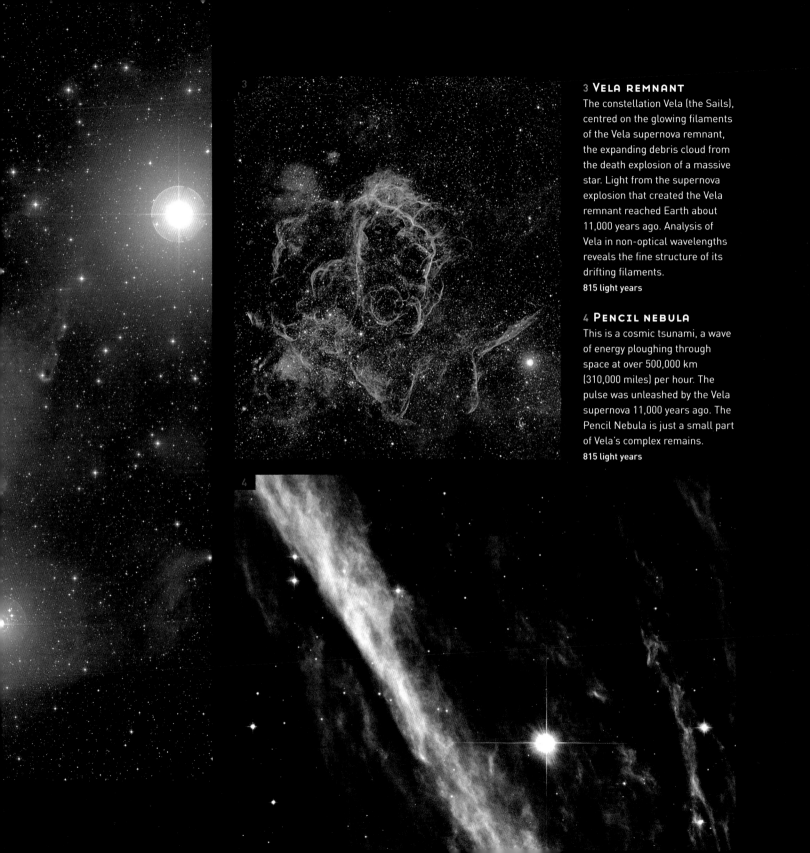

3 Vela remnant

The constellation Vela (the Sails), centred on the glowing filaments of the Vela supernova remnant, the expanding debris cloud from the death explosion of a massive star. Light from the supernova explosion that created the Vela remnant reached Earth about 11,000 years ago. Analysis of Vela in non-optical wavelengths reveals the fine structure of its drifting filaments.

815 light years

4 Pencil nebula

This is a cosmic tsunami, a wave of energy ploughing through space at over 500,000 km (310,000 miles) per hour. The pulse was unleashed by the Vela supernova 11,000 years ago. The Pencil Nebula is just a small part of Vela's complex remains.

815 light years

THE GREAT NEBULAE

Stars are just the brightest and most obvious objects in the night sky. Ghostly-thin regions of gas and dust are just as important as those occasional, blazing concentrations of matter.

If we look towards one of the most prominent constellations in the night sky, the familiar Orion, we can see the Orion Nebula. It is visible as the middle 'star' in the sword of Orion, the three stars located south of Orion's Belt. That small smudge is our feeble perspective on a vast star-forming region 1,500 light years way, and itself part of an even grander system of nebulae. These, in turn, are merely typical examples of nebulae found widely scattered throughout the galaxy.

Like all celestial objects, nebulae are classified according to type. Emission nebulae are clouds of high temperature gas. The atoms in the cloud are energized by ultraviolet light from nearby stars, and the gas emits its own observable radiation as the atoms fall back into lower energy states. These nebulae are usually red because the emission colour of hot hydrogen happens to be red. Green emissions come from oxygen atoms.

Reflection nebulae are exactly what they sound like. They reflect light from stars. Absorption nebulae are in some ways the most interesting and most frustrating. They fascinate us because their relatively dense clouds include dust and complex molecules from which new stars and planets can form. They are frustrating because they often obscure the light from other, equally interesting targets in the distance behind them.

GHOSTLY NEBULA IN CEPHEUS *Often described by observers as a 'dusty curtain' or 'ghostly apparition', reflection nebula vdB 152 is very faint. It lies along the northern Milky Way in the constellation Cepheus. Pockets of dust block light from background stars. Ultraviolet light from the star that happens to be passing through the cloud causes a dim red luminescence in the dust (1,200 light years).*

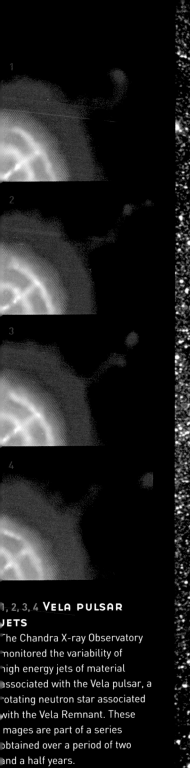

1, 2, 3, 4 VELA PULSAR JETS
The Chandra X-ray Observatory monitored the variability of high energy jets of material associated with the Vela pulsar, a rotating neutron star associated with the Vela Remnant. These images are part of a series obtained over a period of two and a half years.

5 M7 STAR CLUSTER

This prominent cluster, dominated by bright blue stars, can be seen with the naked eye in the tail of the constellation Scorpius. M7 contains about a hundred stars in a region 25 light years across. Also visible is a dark dust cloud near the bottom of the frame, and millions of unrelated stars towards the Galactic centre.

1,000 light years

6 STARDUST IN ARIES

At the lower right of this image, spanning 30 light years, is a dusty blue reflection nebula surrounding the bright star van den Bergh 13 (vdB 13). Dark nebulae sprawl across the scene, hiding nascent stars from optical telescopes.

1,000 light years

1

1, 2 DUMBBELL NEBULA

This planetary nebula is a typical example of a star about the same size as our Sun at the end of its life [1]. The outer layers are being shed into space in processes much slower and less violent than occur in the supernovae of larger stars. Surrounding gas is illuminated by the energetic ejections [2]. The term 'planetary' is a legacy of old-fashioned observational errors.

1,200 light years

2

3 Iris Nebula

Beautiful swirls of gas and dust blossom like flower petals amid the star fields of the constellation Cepheus. Properly catalogued as NGC 7023, the dusty nebular material surrounds a hot, young star. Faint reddish luminescence occurs as some of the dust absorbs the star's invisible ultraviolet radiation and sheds less energetic visible red light. Infrared observations indicate that this nebula may contain complex carbon-based molecules.

1,300 light years

1, 2 **TRAPEZIUM CLUSTER**
The bright stars are well known
as the heart of the Trapezium, an
open cluster of stars in the
centre of the Orion Nebula [1].
The many dim objects, however,
are not well known, and have
come to attention only by the use
of infrared instruments [2].
These are thought to be brown
dwarfs and free-floating planets.
1,300 light years

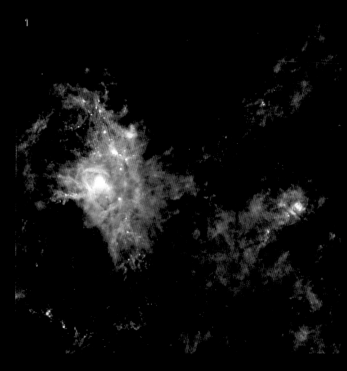

1 GOULD'S BELT

Gould's Belt is a giant ring of stars that circles the night sky. The Solar System just happens to lie near the centre of the belt, which supplies bright stars to many constellations. This detailed infrared view from the Herschel Space Observatory shows a dark cloud of the Belt, located in the Eagle constellation. It covers an area 65 light years across. Some 700 young stars are crowded into the filaments.

1,300 light years

2 NORTH AMERICA NEBULA

We cannot help but find familiar-looking shapes in the night sky. This emission nebula, catalogued as NGC 7,000, and located north-east of bright star Deneb in the constellation of Cygnus, resembles the continent of North America. To the right of the image is the less luminous Pelican Nebula.

1,500 light years

BARNARD 30

Infant stars take shape in the Orion constellation. Astronomers suspect that shockwaves from a supernova explosion three million years ago may have triggered this cycle of star birth. The region is called Barnard 30, and sits on the right side of Orion's 'head', just north of the massive star Lambda Orionis.

1,300 light years

1, 2 THE ORION NEBULA

This great swathe of dust and gas stretches across more than 100 light years, and backdrops most of the Orion constellation [1]. This is one of the closest and most easily studied major star-formation regions. As our astronomical instruments achieve ever greater resolving power, we learn more from Orion about how stars and planets coalesce from collapsing clouds of gas and dust. Formally catalogued as M42, Orion is not so much one cloud, but several, depending on how we classify the different features. It consists of reflection and absorption nebulae, all interacting. Observations at different wavelengths reveal Orion's hidden structures [2], its chemical composition and temperature gradients. It is one of the most intensively studied 'objects' in all of astronomy.

1,500 light years

3 STARS IN ORION

A colony of hot, young stars powers this region of the Orion Nebula, as captured by from NASA's Spitzer Space Telescope. The hottest stars in the region form the Trapezium cluster at centre right.

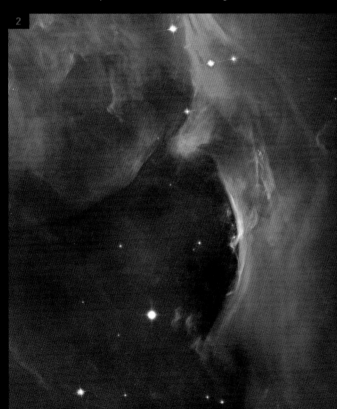

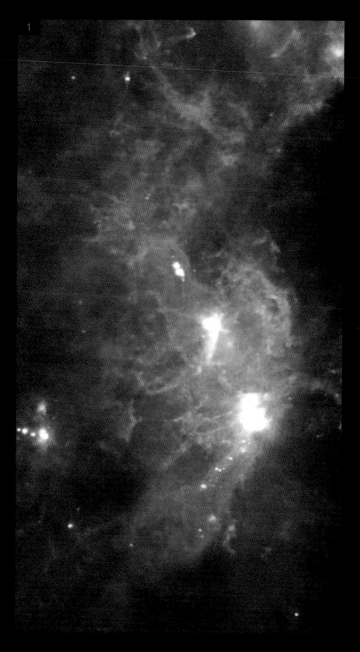

1 ORION STAR FORMATION

Specific Orion regions, observed in infrared, show concentrations of material on the verge of transforming into millions of new stars. The scale of the nebula as a whole defies any human attempt to describe it.

2

2 Santa Claus

A combination of Spitzer infrared
data and x-ray observations,
reveals a huge cloud of high-
temperature gas within Orion.
According to some observers,
this cloud looks like a silhouette
of Santa Claus, but in a blue
rather than red hood.

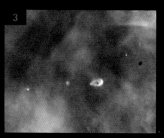

3

3 Orion Proplyds

Evidence is mounting that
planets and solar systems
are common in our galaxy.
Yet another close-up of a region
within the Orion Nebula shows
what seem to be disks of dust
and gas surrounding newly
formed stars. These fuzzy
blobs, called 'proplyds', are
probably infant solar systems in
the process of formation. Of the
six stars visible in these frames,
four appear to have proplyds.
A more complete survey of 110
stars in the region found 56
with proplyds.
1,500 light years

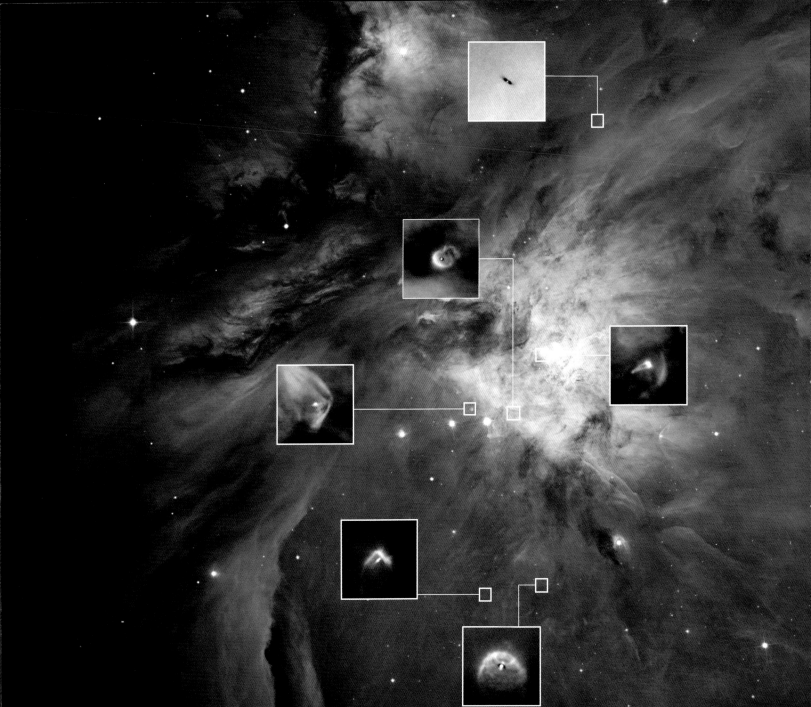

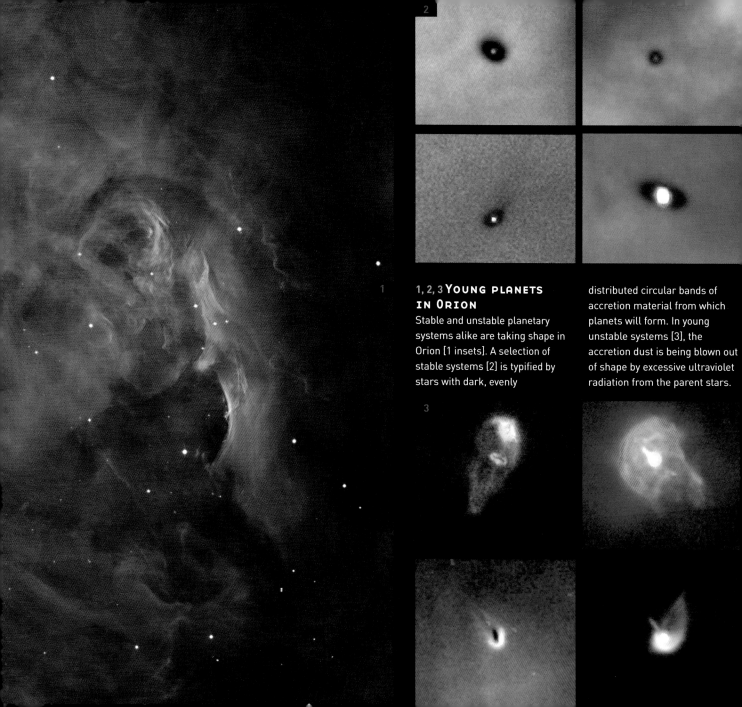

1, 2, 3 YOUNG PLANETS IN ORION

Stable and unstable planetary systems alike are taking shape in Orion [1 insets]. A selection of stable systems [2] is typified by stars with dark, evenly distributed circular bands of accretion material from which planets will form. In young unstable systems [3], the accretion dust is being blown out of shape by excessive ultraviolet radiation from the parent stars.

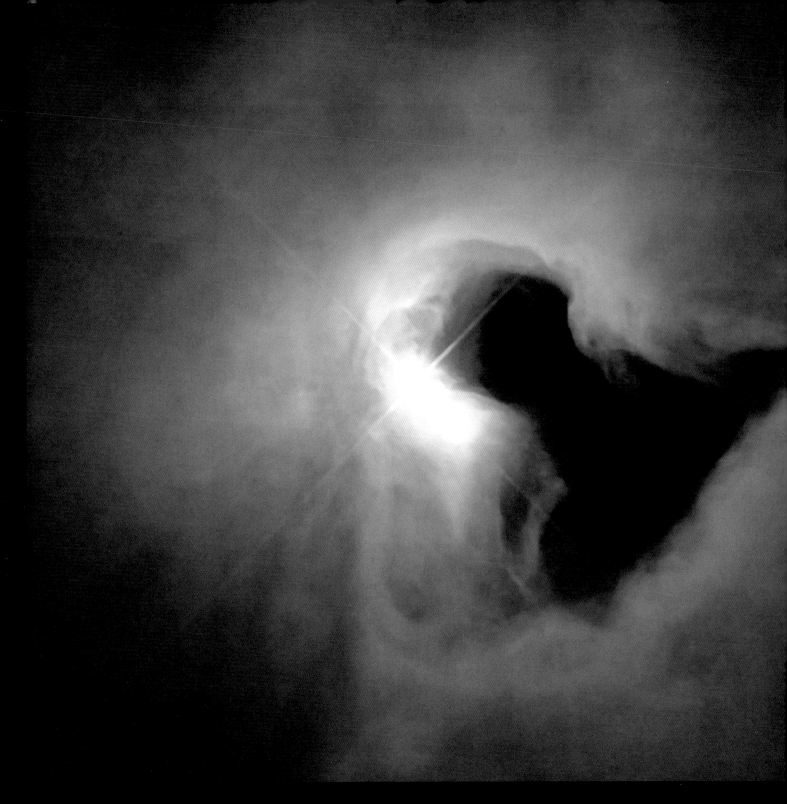

1 NGC 1999
Stellar Cocoon

Just south of Orion's well-known M42 nebula lies this smaller dusty relative, catalogued as NGC 1999. The bright zones of the nebula are lit by a nearby star. The ominous dark area, known as a Bok Globule, is a foreground condensation of cold gas and dust so thick that it blocks light.

1,500 light years

2 Flame nebula

Also known as NGC 2024, this nebula's suggestive red hue colour is causes by hydrogen atoms glowing at the edge of the giant Orion nebula. The hydrogen is being agitated by ultraviolet energy from a young, massive star hidden behind the darker dust.

1,500 light years

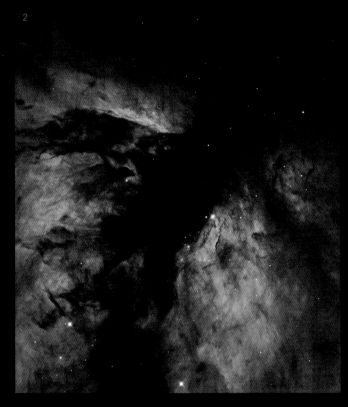

3 LL Orionis bow shock

This young star, moving swiftly through space, is colliding with the dust and gas within the Orion Nebula. Just like a ship ploughing through water, the star creates a bow shock, a wave of compression pushed ahead of it by the star's own intense radiation of energy and subatomic particles.

1,500 light years

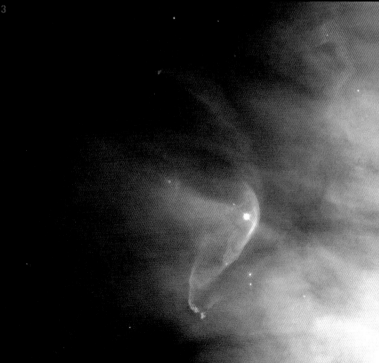

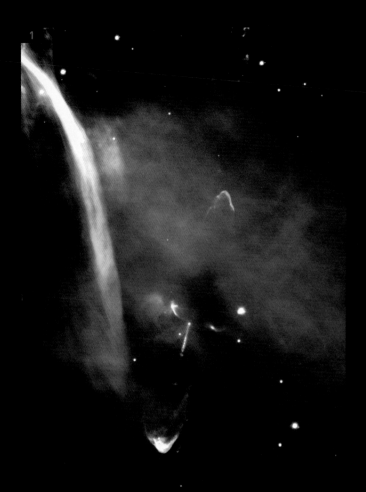

1 HH-34 HERBIG-HARO OBJECT

This young star is spitting out material into space in distinct, interrupted bursts, like 'bullets' of high-energy particles. They appear as red streaks. The bursts may be rebound effects generated whenever gas from the surrounding region collapses onto the star.

1,500 light years

2 FLAMING STAR AE AURIGAE

Stars are often said to 'burn' but this is just shorthand for complex nuclear processes. The smoke-like material seen here is mostly interstellar hydrogen with dark filaments of carbon-rich dust. AE Aurigae is known as a 'runaway' star, because it seems to have been ejected from the Orion Nebula region about 3 million years ago.

1,500 light years

1, 2, 3 HORSEHEAD NEBULA

This famous and distinctly equine-shaped nebula is embedded in the vast and complex Orion Nebula, as seen in this natural-colour image from the Canada-France-Hawaii Telescope in Hawaii [1]. The dark molecular cloud is visible only because its obscuring dust is silhouetted against another, brighter nebula. The prominent horse head portion is part of a larger cloud of dust which can be seen extending toward the bottom of the picture. Closer investigation of the Horsehead's dark absorption core [2,3] shows the extraordinary density of the dust.

1,500 light years

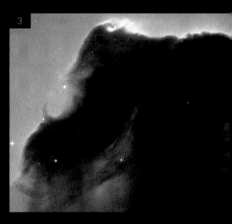

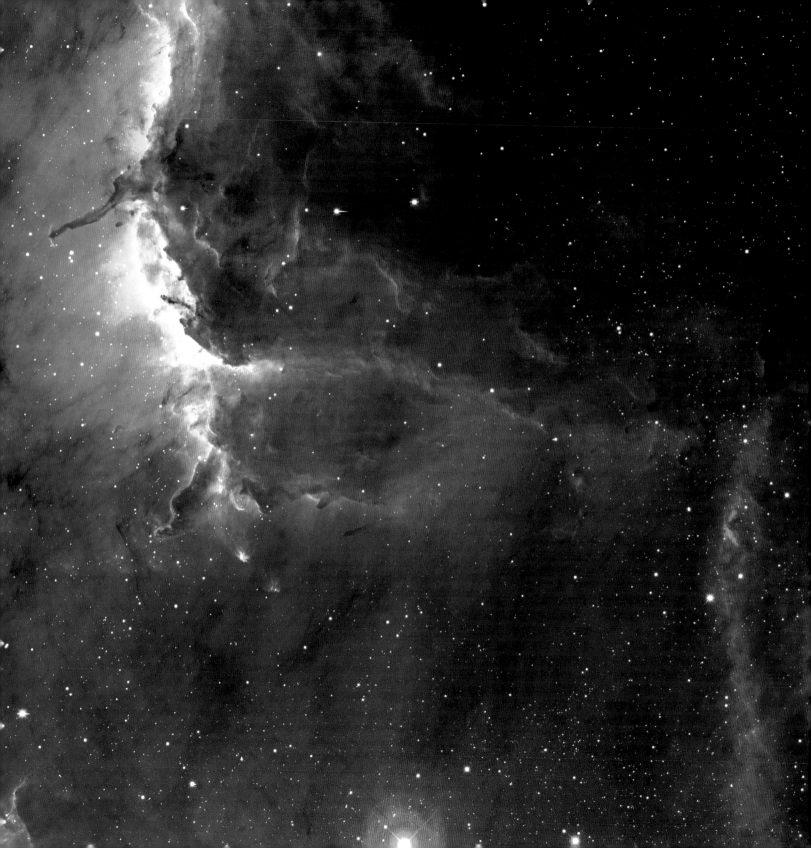

1 Pelican Nebula

This is a particularly active mix of star formation and evolving gas clouds. The light from young energetic stars is slowly transforming cold gas to hot gas, with the advancing boundary between the two known as an 'ionization front'. Most of the bright stars lie off the top of the image, but the bright ionization front is clear.

1,900 light years

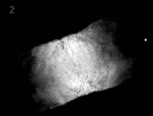

2 Retina Nebula

Our side-on view of this planetary nebula is deceptive. It is actually shaped like a torus, or a donut. The dying star within heats the thick ring of gas and makes it fluoresce.

1,900 light years

3 Eight-burst Nebula

Contrary to what our eyes might tell us, multiple ejections of material from the small, dim star at top right, rather than the more obvious central bright star, created this planetary nebula associated with a binary star system.

2,000 light years

4, 5 M78 Emission Nebula

Columns of dark dust highlight M78 and other bright reflection nebulae in the constellation of Orion [4]. The dark filamentary dust absorbs light from distant stars, while reflecting light from several bright blue stars recently formed within. M78 is on the upper right, while NGC 2071 can be seen to its lower left. Looking at M78 in yet more detail [5], light from newborn stars spills out from depths of the dust core.

1,600 light years

6 Spirograph Nebula

This compact planetary nebula is just one tenth of a light year in diameter. This is still sufficient to encompass our Solar System 100 times over. The granular patterns in the shell of gas that are not well understood. They may be related to chaotic winds from the variable central star, which changes brightness unpredictably every few hours.

2,000 light years

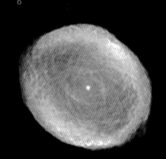

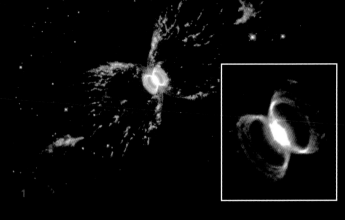

1 SOUTHERN CRAB

This planetary nebula in the constellation Centaurus shrouds two stars, a white dwarf and a red giant, each with a mass equal to the Sun's. The red giant pulsates, shedding material into an accretion disk surrounding the white dwarf, which apparently responds with intermittent thermonuclear explosions.

2,000 light years

3 RING NEBULA

A classic example of a planetary nebula seen as if staring directly down its barrel. Visible even with a small telescope in the constellation of Lyra, it is roughly 500 times the diameter of our Solar System.

2,000 light years

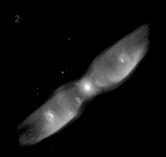

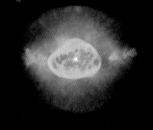

2 TWIN-JET NEBULA

As low-mass stars die, they become white dwarfs by casting off their outer gaseous envelopes. The ejected gas usually forms a display that fades gradually over thousands of years. M2-9, a butterfly-type planetary nebula, is shaped by two stars, one of which is becoming a white dwarf.

2,100 light years

4 BLINKING EYE NEBULA

When viewed through an underpowered telescope, nebula NGC 6826 appears to 'blink' because it is dim. Human eyes have their least sensitive cells toward the centre of the retina. It will even disappear entirely if gazed at too intently.

2,200 light years

5 RED RECTANGLE NEBULA

An aging binary star system towards the constellation of the Unicorn (Monoceros) powers this unusual nebula. A thick torus of dust and gas pinches the flow of materials into two cone-shaped funnels. Because we view these structures edge-on, the boundary edges of the cones seem to form an X shape. The distinct rungs suggest the flows happen in fits and starts.

2,300 light years

6

6 IC1396 IN CEPHEUS

This emission nebula is a mix of
glowing cosmic gas and dark
dust clouds in the constellation
of Cepheus. Energized by the
bright, bluish central star, this
star forming region sprawls
across hundreds of light years,
spanning over three degrees of
the night sky. The Elephant's
Trunk nebula lies just below
centre.

3,000 light years

1

1, 2 ELEPHANT'S TRUNK NEBULA

This nebula winds through the young star cluster complex IC 1396 in the constellation of Cepheus. The 'trunk' is over 20 light years long. This image [1] was recorded through filters that show light from ionized hydrogen, sulphur and oxygen atoms in the region. NASA's Spitzer Space Telescope captured a glowing stellar nursery within a dark globule of dust that is opaque to visible light [2].

3,000 light years

2

3, 4, 5 PROTOSTARS IN IC 1396

Within this globule of the IC1396 nebula [3], a half-dozen newly discovered protostars, or embryonic stars [4,5] become discernible in non-optical infrared wavelengths.

3,000 light years

6 NGC 217

A dusty nebula in the constellation Monoceros (Unicorn) shines at the upper left. Reflecting the light of nearby hot stars, it is joined by other bluish reflection nebulae and a compact red emission region.

2,400 light years

7 HUBBLE'S VARIABLE NEBULA

Dense knots of gas partially obscuring the Monoceros constellation cast strange shadows. The light comes from the star R Monocerotis. This nebula noticeably changes its appearance over just a few weeks.

2,500 light years

1, 2 **CHRISTMAS TREE CLUSTER**

An open cluster of about 40 stars in the constellation Monoceros forms the outline of an inverted fir tree [1] in an array of several conjoined nebulae, including the distinct Cone Nebula at bottom, and revealed in this more detailed image [2]. The star S Monocerotis (also known as 15 Mon) at the base of the 'tree' is about 8,500 times as luminous as the Sun.

2,500 light years

1

2

3, 4, 5, 6 Veil Nebula

These wisps of gas [3,4] are all that remain of NGC 6960, a Milky Way star in the constellation of Cygnus that no longer exists. Many thousands of years ago the star exploded in a supernova. At the time, the expanding cloud would have been as bright in our night sky as a crescent Moon, and visible for many weeks to our ancestors. This delicate detail of the nebula, as revealed by the Hubble Telescope [5] is 100 light years across. a wider field of view [6] shows the unrelated star 52 Cygni, seen in bright blue, apparently riding the nebula like a witch on a broom.

2,600 light years

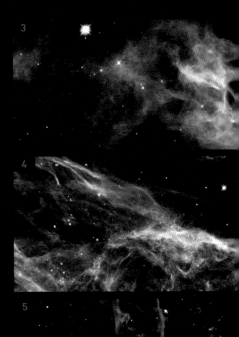

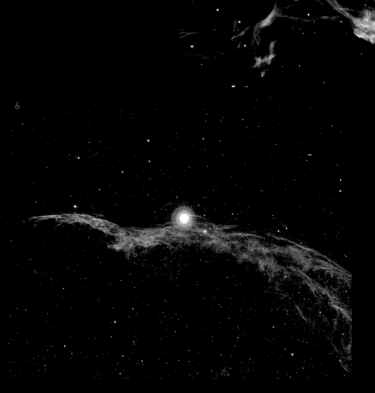

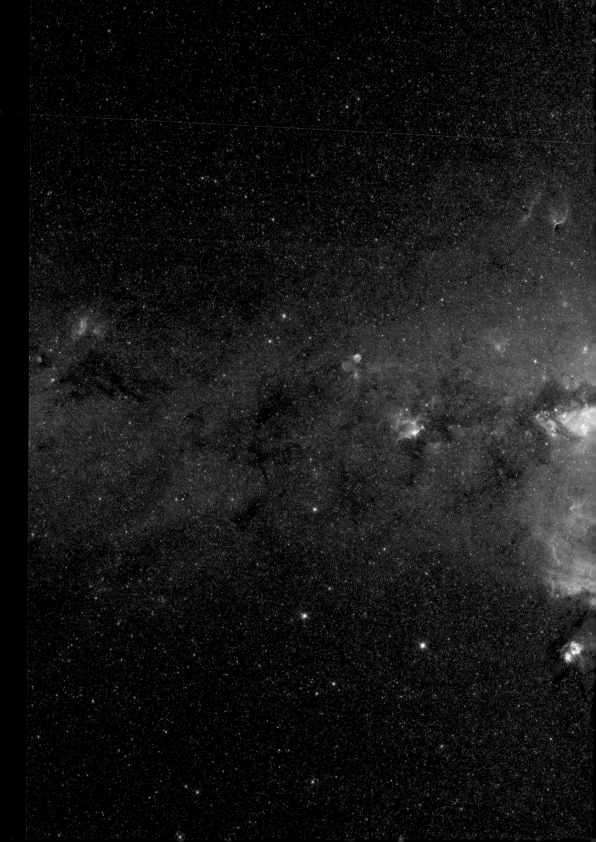

Second-Brightest Star

Dubbed the 'Peony nebula' star, this blazing ball of gas in the constellation Sagittarius shines with the equivalent light of 3.2 million Suns. The only star that outshines it, Eta Carina in the Carina constellation, produces the equivalent of 4.7 million Suns, but the question is still debated, and the Peony nebula star could be the brightest yet seen in our galaxy.

2,600 light years

INTO THE MILKY WAY

The swathe of stars that we see in the night sky and call the 'Milky Way' is just a very small nearby fraction of our galaxy, seen edge-on from our perspective near the galaxy's edge. Modern telescopes take us on a journey into its core.

It's something of a myth that our Solar System lives on the outermost edges of the galaxy, as if haunting the unfashionable suburbs of a great stellar city. In fact we are about halfway out from the centre, on the inner edge of the Orion–Cygnus Arm. The Sun, together with all its planets, moons, asteroids and other bodies, orbits the galactic centre at about once every 220 million years.

The Milky Way's core bulges above and below the galactic plane of rotation for a thickness of more than 3,000 light years. The central bulge is elongated, and probably about 32,000 light years across at is widest point. In our neighborhood, stars are separated by distances ranging from about five to ten light years. A cubic light year of space contains about 700 stars. As we head toward the galactic centre, the population density is orders of magnitude greater, with hundreds of thousands of stars crowded into volumes just a few light years across. Even so, the distances between stars remain unimaginably vast in human terms.

GOULD'S BELT *First described by Benjamin Gould in 1879, this region encompasses several molecular clouds in a local part of our galaxy where star formation is prominent. The Belt is relatively close to our Solar System, and the Galactic Centre is approximately 16 times more distant (3,000 light years).*

4 LL PEGASI

This strange spiral is probably related to a dying star in a binary system that is shedding its outer atmosphere. The huge spiral spans about a third of a light year across and, winding four or five complete turns, has an exceptionally neat regularity.

3,000 light years

1 CIRCINUS MOLECULAR CLOUD COMPLEX

The Wide-Field Infrared Survey Explorer, or WISE, uncovered a striking population of young stellar objects in a complex of dense, dark clouds in the southern constellation of Circinus. This mosaic covers an area on the sky so large that it could contain 800 full Moons. The cloud itself is more than 180 light-years across.

2,280 light years

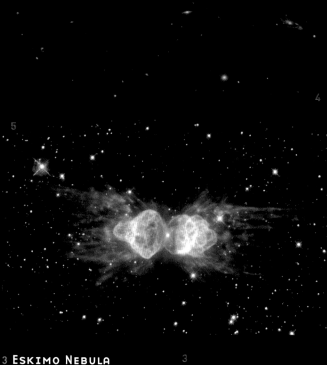

5 ANT NEBULA

This nebula challenges old ideas about what eventually happens to stars similar to our own Sun. Their fate is more complex and dramatic than astronomers previously believed. Although the ejection of gas from the dying star in the Ant Nebula is violent, it is unusually symmetrical.

3,000 light years

2 EGG NEBULA

Thick dust blocks the centre star from view, while the dust shells further out reflect its light. The dust acts as a polarizing filter, adjusting the planes of the lightwave oscillations and creating this pattern, false-colour coded to highlight the orientation of polarization.

3,000 light years

3 ESKIMO NEBULA

In 787AD, astronomer William Herschel discovered the Eskimo Nebula. From the ground, the nebula, catalogued as NGC 2392, resembles a person's head surrounded by a hood. In 2000, the Hubble Space Telescope resolved more of its complex details, shaped by the outer layers of a Sun-like star that were shed only 10,000 years ago.

3,000 light years

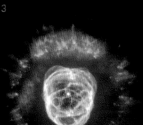

6 M35 OPEN CLUSTER →

Open clusters may contain from 100 to 10,000 stars, all of which formed at nearly the same time. M35 is relatively young at 150 million years old, and diffuse, with about 2,500 stars scattered over a volume 30 light years across. A more compact cluster, NGC 2158, four times more distant and ten times older, is visible on the lower right. Both clusters are visible near the constellation of Gemini.

2,800 light years

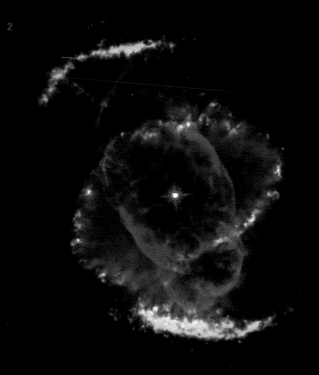

4 BUTTERFLY →

This celestial object, located in Scorpius, looks like a delicate winged insect, but it is far from serene. The 'wings' are roiling cauldrons of hot gas tearing across space so fast, it could travel from the Earth to the Moon in just a few minutes. A dying star that was once about five times the mass of the Sun is at the centre of this fury.

3.8 thousand light years

1, 2 CAT'S EYE NEBULA

A dying star throws off shells of glowing gas [1]. This Hubble Space Telescope image reveals the Cat's Eye as one of the most complex and little-understood planetary nebulae known. The bright central object is probably a binary star system [2].

3,000 light years

3 NGC 7129 AND NGC 7142

An intriguing visual pairing of dusty reflection nebula NGC 7129 (left) and open star cluster NGC 7142. They appear separated by only half a degree in the sky, but actually lie at different distances. NGC 7129 is 3,000 light years away, while NGC 7142 is twice that far away. Foreground dust reddens the light from the distant cluster.

3,000 light years

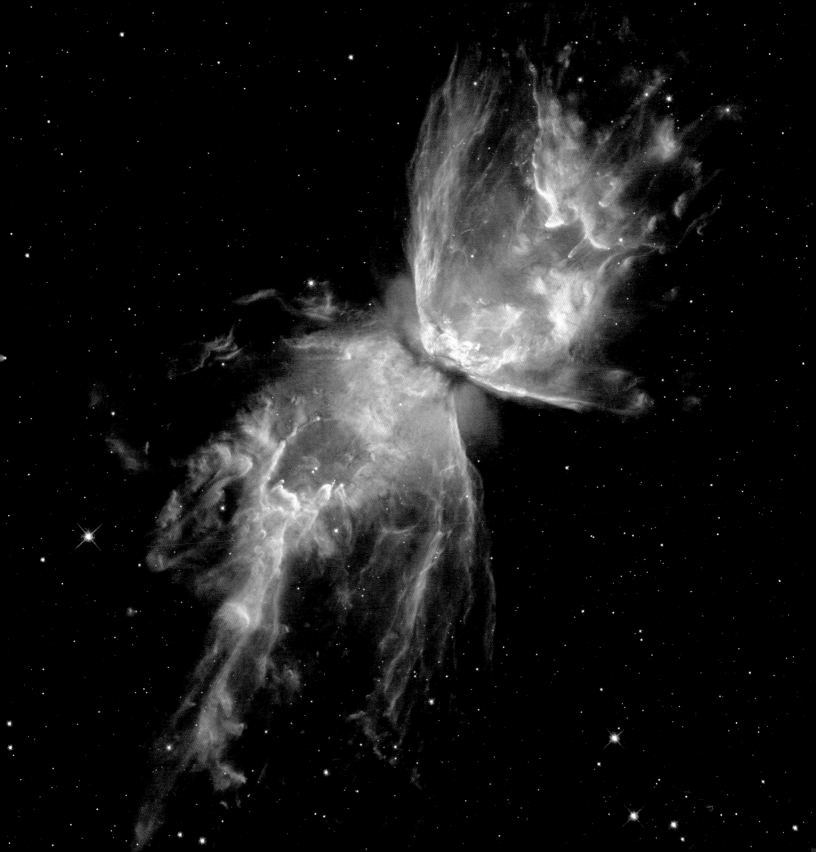

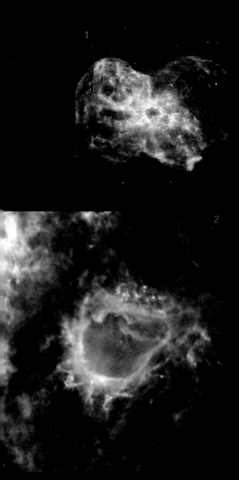

1 Chaotic planetary nebula

Another butterfly-shaped planetary nebula, designated NGC 2440, contains one of the hottest white dwarf stars known. A multi-spectral scan reveals more of the nebula's structure. In approximately 5 billion years, our own Sun will look like this.

4,000 light years

2 RCW 120 Bubble Region

This bubble contains an embryonic star ten times heavier than our Sun. Surrounding it is enough gaseous material to build 2,000 more stars. It pushes on the surrounding dust and gas with nothing more than the power of its starlight. In the 2.5 million years it has existed, it has raised the density of materials in the bubble walls. Some of it will collapse to form yet more stars.

4,300 light years

3 Bug Nebula

Sometimes confusingly called the Butterfly Nebula, even though other objects of that same name exists, it is catalogued as NGC 6302 in the constellation Scorpius. The 'wings' consist of superheated gas tearing across space fast enough to travel from Earth to the Moon in half an hour. A dying star about five times the mass of the Sun is at the centre of this fury.

4,000 light years

4 Bubbles and Jets

The Hubble Space Telescope found a wealth of previously unseen structures in planetary nebula NGC 2371 in the Gemini constellation. The remnant star visible at the centre is the super-hot core of a former red giant, now stripped of its outer layers. Prominent pink clouds of cool, dense gas lie on opposite sides of the central star. Also striking are the numerous, very small pink dots, marking small, dense knots of gas.

4,300 light years

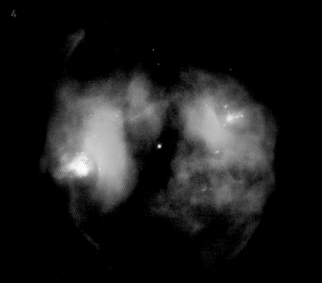

5 White Dwarf Remnant

Another butterfly-shaped planetary nebula, designated NGC 2440, contains one of the hottest white dwarf stars known. It can be seen as the bright dot near the photo's centre. In approximately 5 billion years, our own Sun will look like this.

4,000 light years

6 Red Spider Nebula

Huge waves are sculpted in this two-lobed nebula in the constellation of Sagittarius, which harbours an extremely hot star. Its radiant winds of energy and subatomic matter generate waves 100 billion km (62 billion miles) high. The waves are caused by supersonic shocks as the surrounding gas is compressed.

4,000 light years

1, 2 MAVERICK STARS

This montage of Hubble images [1] shows four "runaway" stars, each just a few million years old. The gravitational influences of other stars have cast them adrift from their original star-formation regions. They are moving at more than 180,000 kmph (112,000 miles per hour) and creating bow shocks in surrounding gas clouds. A much closer runaway, Zeta Ophiuchi, the bright blue star at the centre of this image [2] is a hot star twenty times larger than the Sun. It is just 458 light years way, a proximity that enables us to study its bow shock in detail.

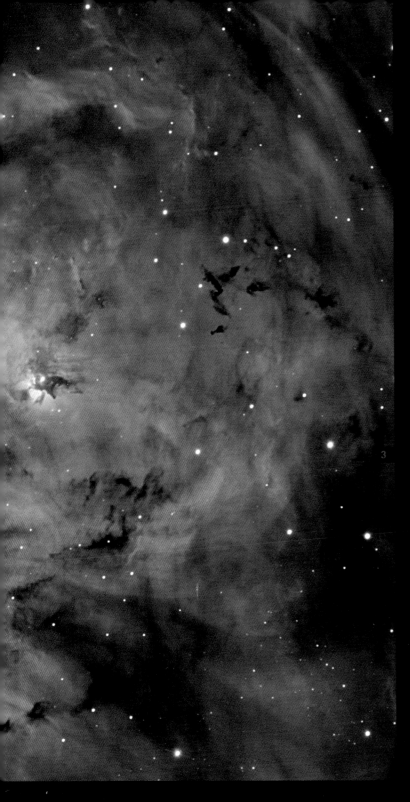

3, 4, 5 EASILY VISIBLE NEBULA

The impressive Lagoon Nebula is home for many young stars and hot gas. Spanning 100 light years across, yet comparatively close to our region of the Milky Way, this Nebula [3] is so large and bright that it can be seen, even without a telescope, toward the constellation of Sagittarius. Resolved in more detail [4] the nebula contains a number of Bok Globules – dark, collapsing clouds of protostellar material. It also includes a funnel-like or tornado-like structure shaped by the energy of a hot star's ultraviolet light. Closer still, and details of the nebula's swirling gas are resolved [5].

5,000 light years

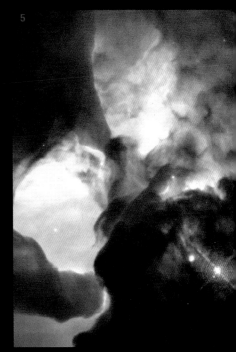

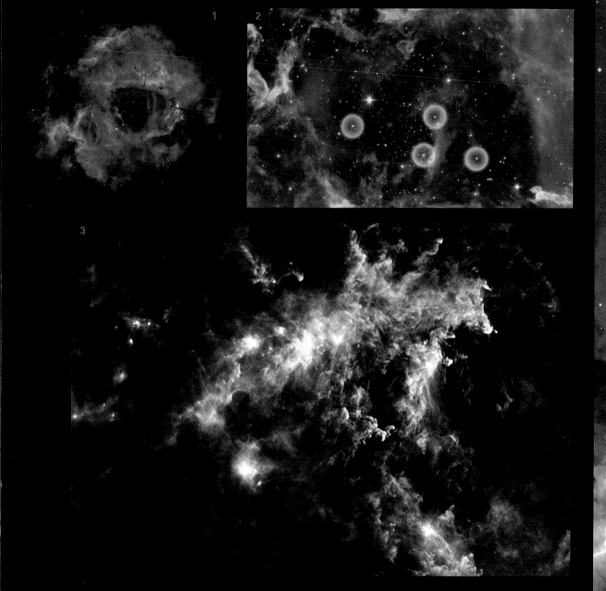

1, 2, 3, 4 DANGEROUS BLOOM

Lurking inside the innocent-looking cosmic flower of the Rosette Nebula in Monoceros [1] and in closer detail in this image [2] are planetary danger zones surrounding super-hot stars, called O-stars, which give off intense winds and radiation. Young, cooler stars that stray too close are in danger of having their dusty planet-forming materials stripped away. This infrared image [3] is overlaid with a graphic showing the danger zones, and this detailed view from the Herschel Space Observatory [4] shows some the dense clouds associated with the nebula.

5,000 light years

1 Coldest Place in Space

This young planetary nebula in the constellation of Centaurus, known as the Boomerang Nebula, is one of the universe's most peculiar places, the coldest region found so far, only fractionally warmer than absolute zero, the lowest limit for all temperatures. The rapid expansion of the nebula has caused its intense plunge in temperature.

5,000 light years

4 Cat's Paw Nebula →

The Hubble Space Telescope may be among the most famous astronomical instruments, by ground-based telescopes also deliver impressive results, as in this photograph of the Cat's Paw Nebula in Sciorpius obtained by the Blanco 4-metre Telescope in Chile.

5,500 light years

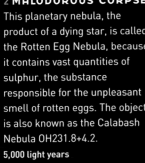

3 About to blow

Massive stars lead short, spectacular lives. This composite x-ray image reveals dramatic details of a portion of the Crescent Nebula, a gaseous shell created by powerful winds from the massive star HD 192163, out of frame to the lower right. This star will become a supernova in about a hundred thousand years' time. The force of its wayward energies have created two shock waves, one moving outward to create the green filamentary structure, and one compressing inward to produce a bubble of million degree Celsius x-ray emitting gas, seen in blue.

5,000 light years

2 Malodorous corpse

This planetary nebula, the product of a dying star, is called the Rotten Egg Nebula, because it contains vast quantities of sulphur, the substance responsible for the unpleasant smell of rotten eggs. The object is also known as the Calabash Nebula OH231.8+4.2.

5,000 light years

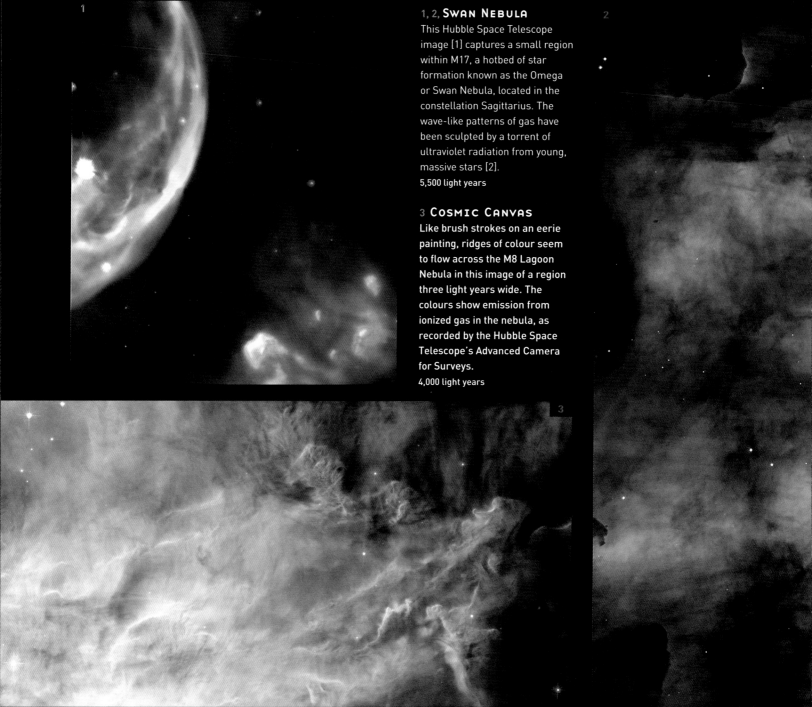

1, 2, SWAN NEBULA

This Hubble Space Telescope image [1] captures a small region within M17, a hotbed of star formation known as the Omega or Swan Nebula, located in the constellation Sagittarius. The wave-like patterns of gas have been sculpted by a torrent of ultraviolet radiation from young, massive stars [2].

5,500 light years

3 COSMIC CANVAS

Like brush strokes on an eerie painting, ridges of colour seem to flow across the M8 Lagoon Nebula in this image of a region three light years wide. The colours show emission from ionized gas in the nebula, as recorded by the Hubble Space Telescope's Advanced Camera for Surveys.

4,000 light years

1, 2 NGC 3576
Sculpted Arcs

These alluring whorls of gas and dust [1] drift through the Sagittarius arm of our spiral Milky Way Galaxy. Within the region, star formation contributes to the complex and suggestive shapes [2]. Powerful winds from young, massive stars within the nebula have sculpted the looping filaments.

5,900 light years

1 BOK GLOBULES

Strange dark clouds known as Bok Globules float serenely in this beautiful image taken with the Hubble Space Telescope. The dense, opaque cloud are silhouetted against nearby bright stars in the busy star-forming region, IC 2944. Globules like these have been known since Dutch-American astronomer Bart Bok first drew attention to them in 1947.

5,900 light years

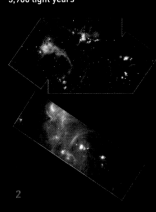

2, 3, FURIOUS NURSERY

Hidden behind a shroud of dust in the constellation Cygnus is a stellar nursery called DR21, which is giving birth to some of the most massive stars in our galaxy. Visible light images [2] reveal no trace of this interstellar cauldron because of heavy dust obscuration. New images from NASA's Spitzer Space Telescope allow us to see behind the veil and pinpoint one of the most massive young stars yet seen in our Milky Way galaxy, some 100,000 times as bright as the Sun. Also revealed for the first time is a powerful outflow of hot gas emanating from this star and bursting through a giant molecular cloud. This colourful composite [3] is assembled from a variety of different wavelengths.

6,000 light years

5

4, 5, 6 HEART AND SOUL NEBULAE

An infrared mosaic from NASA's Wide-field Infrared Survey Explorer, or WISE [4] covers an area of the sky ten times as wide as the full Moon in the constellation Cassiopeia. The Heart and Soul nebulae form a vast star-forming complex in the Perseus spiral arm of the Milky Way. The nebula to the right is the Heart, designated IC 1805. To the left is the Soul Nebula, also known as the Embryo nebula, IC 1848. An optical image of the Heart Nebula [5] reveals its anatomically suggestive shape. Closer details of the

6

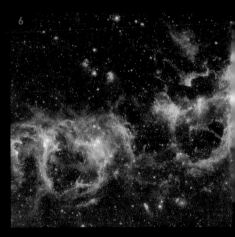

Heart [6] show light emission from hydrogen in green, and oxygen in blue hues.

6,000 light years

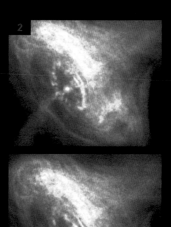

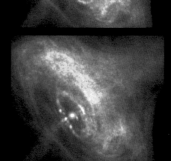

1 CRAB NEBULA

A star's spectacular death in the constellation Taurus was observed on Earth as the supernova of 1054 AD. Now, almost a thousand years later, a superdense neutron star left behind by the stellar death is spewing out a blizzard of extremely high-energy particles into the expanding debris field known as the Crab Nebula.

6,500 light years

2 CRAB PULSAR

The Chandra X-ray Telescope images in this collage were made over a span of several months. They provide a stunning view of the activity in the inner region around the Crab Nebula pulsar, a rapidly rotating neutron star seen as a bright white dot near the centre of the images. Enormous electrical voltages generated by the rotating, highly magnetized neutron star accelerate particles outward along its equator. These voltages also produce the jets seen spewing x-ray-emitting matter and antimatter particles perpendicular to the rings.

6,500 light years

3 CONFUSING PERSPECTIVES

The emission nebula NGC 3603 at the left is 20,000 light years away, while its apparent neighbour, NGC 3582 is a third of that distance away from us. Closer inspection reveals a young star cluster within the more distant nebula.

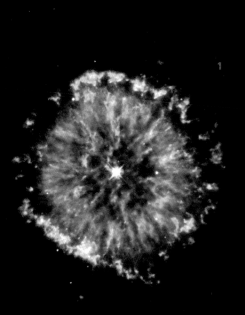

1 NGC 6751

The crush of gravity on a white dwarf is so extreme, it even compresses individual atoms, forcing electrons into the nuclei. This composite colour Hubble image of NGC 6751 is a beautiful example of a classic planetary nebula with complex features. It is in the constellation Aquila.

6,500 light years

2 EAGLE NEBULA'S FAIRY

Dark dust makes the shape of a winged sprite or fairy, but not the kind you might find at the bottom of your garden. This one's ten light years tall.

6,500 light years

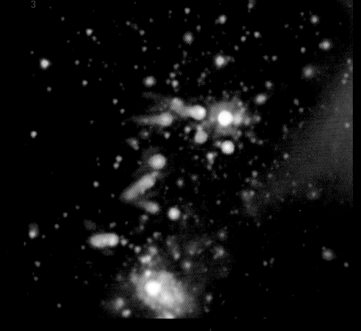

5 Super-supernova

White dwarfs can sometimes be temporarily reanimated, like stellar vampires, if an adjacent star drifts too close, inadvertently contributing fresh fuel to its supposedly depleted predator. A fresh cycle of nuclear fusion re-ignites the star, producing an especially powerful supernova that leaves nothing in its wake except an echoing shell of gas, as in this example, SNR 1572. Its final explosion was observed by Tycho Brahe in 1572.

7,500 light years

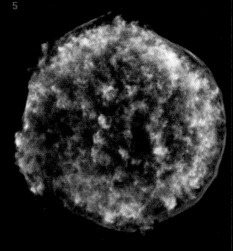

3 Unsuitable for worlds

A portion of the W5 star-forming region, located 6,500 light-years away in the constellation Cassiopeia shows the unpleasant environment near a group of massive stars, where radiation and stellar winds blast any planet-making material away from less aggressive stars. The planetary material can be seen as comet-like tails behind three stars near the centre of the picture.

6,500 light years

4 The brightest burst

This false-colour Chandra image of supernova remnant SN 106 in Lupus shows x-rays produced by high-energy particles (blue) and multimillion degree gas (red/green). In 1006 AD, what was thought to be a new star suddenly appeared in the sky, and over the course of just a few days became brighter than the planet Venus. This may have been the brightest supernova on record.

7,000 light years

6 Closest Cluster

Globular clusters are groups of stars that convene into roughly spherical collections of tightly-spaced stars. This is the closest example, M4, a cluster of 100,000 stars a fraction west of the bright star Antares in the constellation Scorpius.

7,000 light years

1, 2 NGC 6397 Cluster

Another nearby globular cluster [1] sits in the constellation Ara. Its massive stars have exhausted their supplies of nuclear fuel. This should leave the cluster with only old, faint stars, but some younger 'blue stragglers' appear to have joined the cluster as a result of stellar collisions. In this more detailed view [2] the blue squares pinpoint the young white dwarfs and the red circles outline the older white dwarfs.

8,200 light years

THE PILLARS OF CREATION

New stars and planets are born from the dust and gas hurled out during the destruction of other stars. Destruction and creation walk hand in hand throughout the galaxy.

Stars substantially larger than our Sun end their days in a violent cataclysm, during which the outer parts of the star are hurled away in a violent explosion, a Supernova. Over immense spans of time, those gaseous fragments of wreckage accumulate into larger nebulae, and the cycle of birth and death begins anew. The most famous picture from the Hubble Space Telescope, taken in 1995 and now familiar to many millions of people the world over, is dubbed 'The Pillars of Creation'. A detail of the Eagle Nebula in the constellation Serpens shows tall towers, each long enough to stretch from our Sun to the next nearest star, Proxima Centauri. We can clearly see bright stars sprinkled across the image and embedded in the great columns of gas and dust.

The swirls of glowing gas and dark dust in a nebula may look beguilingly substantial, but if you extracted all the air from every room in your house and substituted a house-sized volume of gas and dust scooped from the tip of a Pillar of Creation', you could collect all the material on the tip of a pencil. On the human scale, being inside a nebula is indistinguishable from being inside nothing. On the grand scale of a galaxy the story is different. Vast volumes of seemingly empty space contain countless trillions of tons of material.

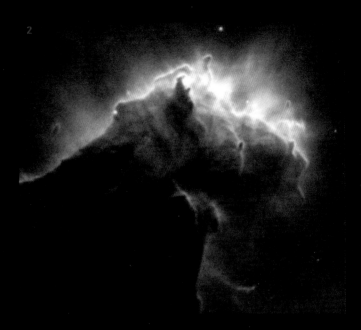

1 INSIDE THE EAGLE NEBULA

The bright region is a window into the centre of a larger dark shell of dust. Through this window, a brightly-lit region appears where an open cluster of stars is taking shape. Tall pillars and round globules of dark dust and cold gas remain where stars are still forming.

2 TIP OF A PILLAR

A closer view at the tip of a 'pillar' shows cool molecular hydrogen gas and dust embedded inside finger-like protrusions extending from the top of the nebula. Each tip is larger than our own solar system.

1, 2, 3, 4, 5 Carina Nebula

This turbulent stellar nursery [1] is named after its location in the southern constellation of Carina. Scorching radiation and fast winds, streams of charged particles from super-hot newborn stars in the nebula [2, 3] shape pillars of dust [4] into random but significant concentrations. Two huge pillars within Carina [5] seem very solid, but a powerful star, not itself visible through the opaque dust, is ejecting powerful blasts of radiation that will eventually dissipate the clouds.

7,500 light years

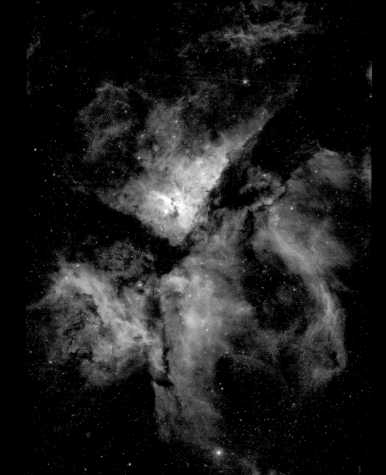

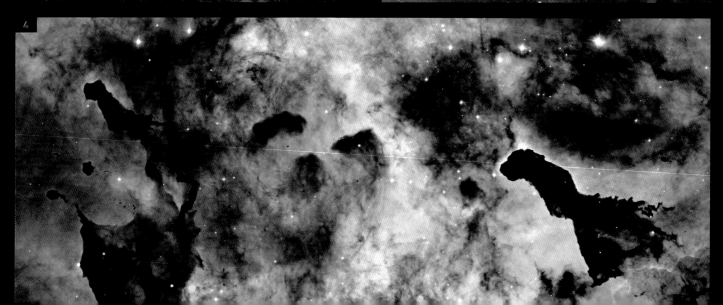

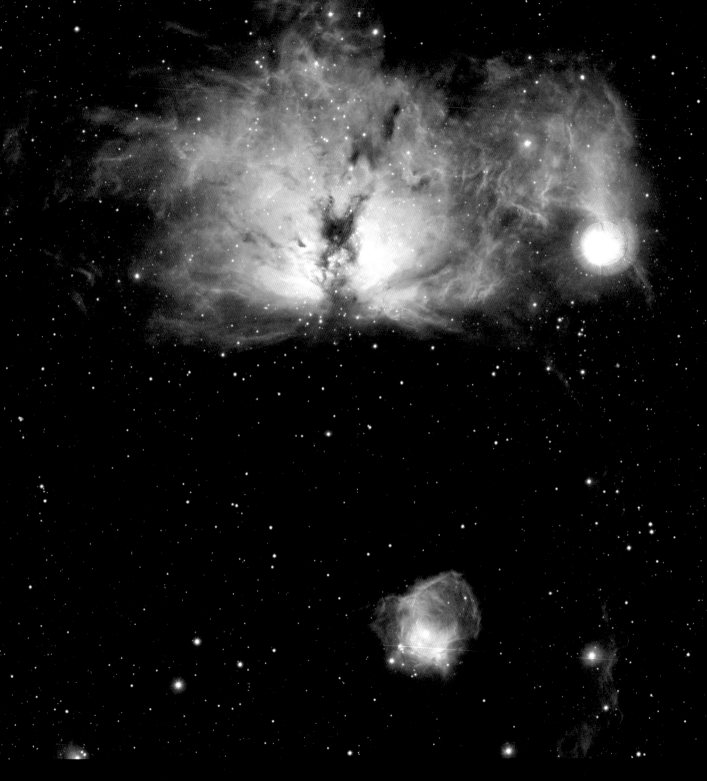

1, 2, 3 CARINA NEBULA

The brightest nebulae, such as the enormous star-forming region of Carina [1], are made to glow by a similar process that activates a fluorescent light, except that it's not electricity so much as ultraviolet radiation from the stars within the nebulae [2] that generates the necessary energy, 'exciting' atoms and molecules within the cloud and raising their energy levels. They swiftly fall back into their normal states, re-releasing the excess energy as visible light. Then more ultraviolet energy is absorbed and the process starts anew. This detailed Hubble image [3] shows a tottering pillar of star-forming gas and dust within the nebula.

8,000 light years

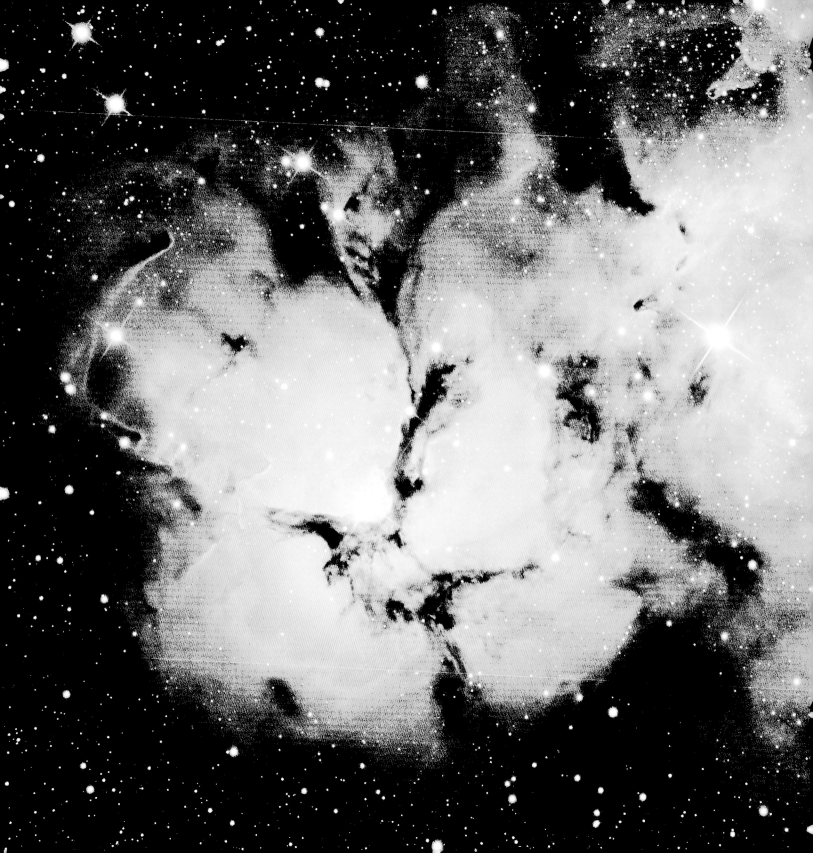

1, 2, 3, 4, 5 TRIFID NEBULA
Also known as M20, this photogenic nebula [1] is visible with good binoculars towards the constellation of Sagittarius. Infrared analysis by the Spitzer telescope [2] reveals a different side to the Nebula. Where dark lanes of dust [3] are visible at optical wavelengths, bright regions of star-forming activity are seen in the Spitzer pictures. This seemingly red-hot region [4] actually picks out the coolest areas of the nebula. Colours in scientific images often do not match our common-sense assumptions. Another view of the dust lanes from Hubble [5] is slightly easier to interpret.

7,600 light years

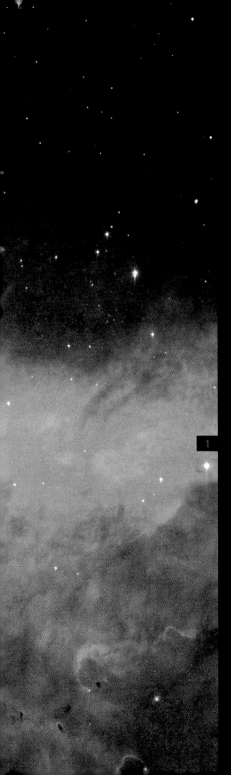

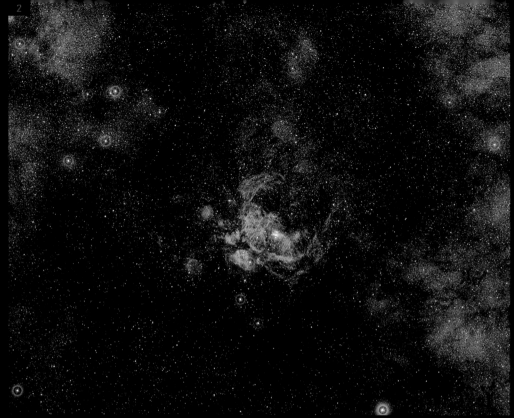

1, 2 WAR AND PEACE NEBULA

No one knows why the star-forming region NGC 6357, known as the Lobster or the War and Peace Nebula [1] is forming some of the most massive stars ever discovered. Near the more obvious Cat's Paw nebula, NGC 6357 houses the open star cluster Pismis 24, home to these tremendously bright and blue stars [2].

8,000 light years

3 HOURGLASS NEBULA

The symmetry of this planetary nebula may be the result of ejected material being shaped by strong magnetic fields, or perhaps by gravitational interaction with another as-yet unidentified object.

8,000 light years

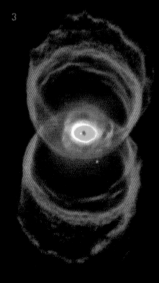

1, 2 PISMIS-24 CLUSTER

At least one star in the open
cluster Pismis 24 is over 200
times the mass of our Sun. The
cluster [1] derives its brilliance
not from a single star but from
three at least. The brightest star,
Pismis 24-1, is a double star [2].
Of these, one could be double
again,making it a double or even
a triple system.

8,000 light years

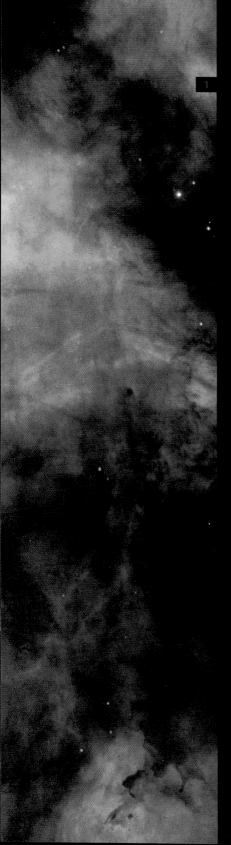

1, 2, 3, 4 KEYHOLE NEBULA

Catalogued as GC 6357, this is a diffuse nebula near NGC 6334 in the constellation Scorpius [1]. Cutting across a nearby star-forming region are 'hills and valleys' of gas and dust displayed in these intricate details. A region near the Keyhole Nebula [3] is haunted by a lonely Bok Globule of dense star-forming material [4]. This one is known as the 'Caterpillar'. Pulling back again, it is worth remembering how few of these details we see in most Earth-based telescopes [2].

8,000 light years

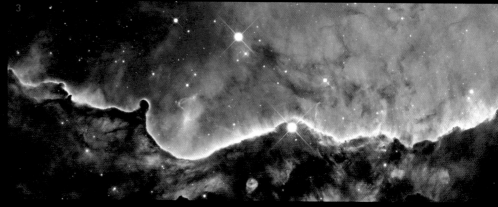

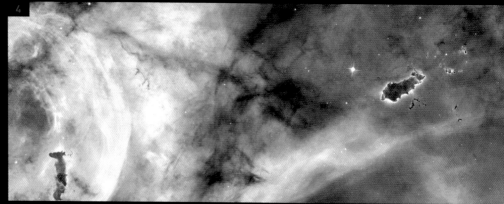

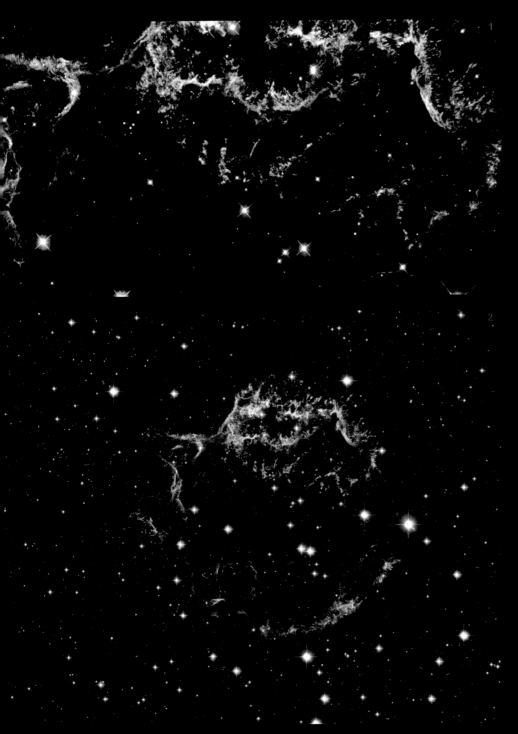

1, 2, 3, 4 LIFE FROM DEATH
We have seen many images of star wreckage, but perhaps none is as beautiful as Cassiopeia A, the debris from a supernova whose brilliant burst of light reached Earth 325 years ago. Details of the wreckage in optical light [1,2] is barely even half the story. A composite of x-ray, optical, and infrared light exposures [4] shows the richness of a catastrophe that eventually will feed essential chemicals, including oxygen, sulphur, silicon and iron, into the birth of new stars and planets, and perhaps of new life. Two processed images taken one year apart by the Spitzer telescope and combined [3] reveal dust and gas features that have not changed over time in grey, and those that have changed in blue or orange. Obviously the ghost of Cassiopeia A is changing fast.
10,000 light years

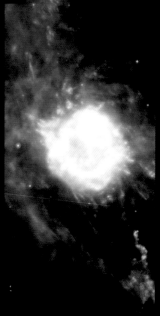

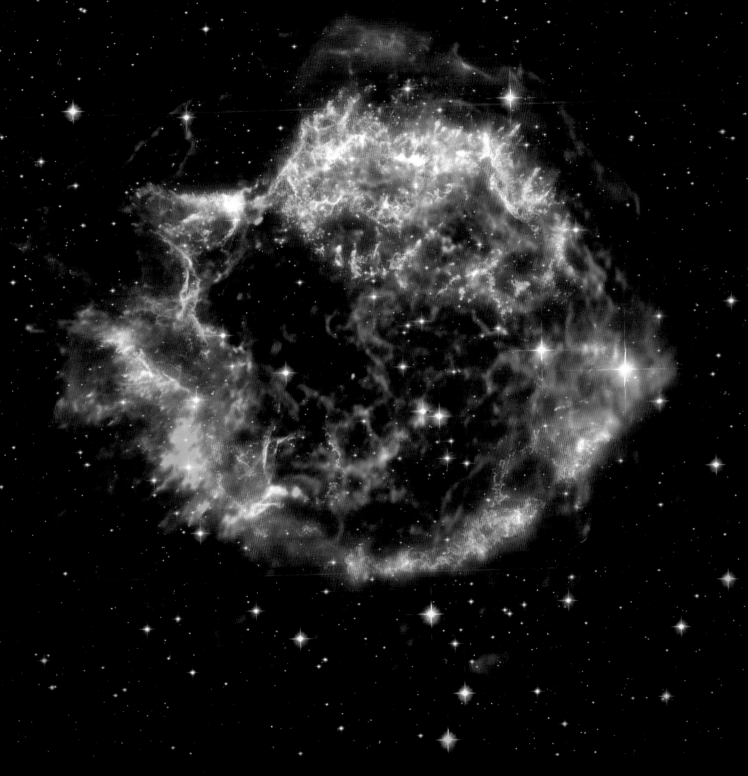

1, 2 STELLAR WORKSHOP

NGC 281 in Cassiopeia is a busy
sector of star formation [2].
Prominent features include a
small open cluster of stars, a
diffuse red-glowing emission
nebula, large lanes of obscuring
gas and dust, and dense knots of
dust and gas in which stars may
still be forming. The open cluster
of stars visible around the centre
has formed only in the last few
million years. Light-absorbing
Bok Globules, knots of dust in
NGHC 281 [1] often frustrate
astronomers trying to see
objects far behind them.

10,000 light years

3 SORTING STARS FROM DUST

Several telescopes worked in
combination to deliver this image
that distinguishes the star IRAS
13481-6124, in the constellation
Centaurus, from its surrounding
disc of cosmic dust. Astronomers
still cannot decide if very large
stars begin their lives in the
same way as small stars.

10,000 light years

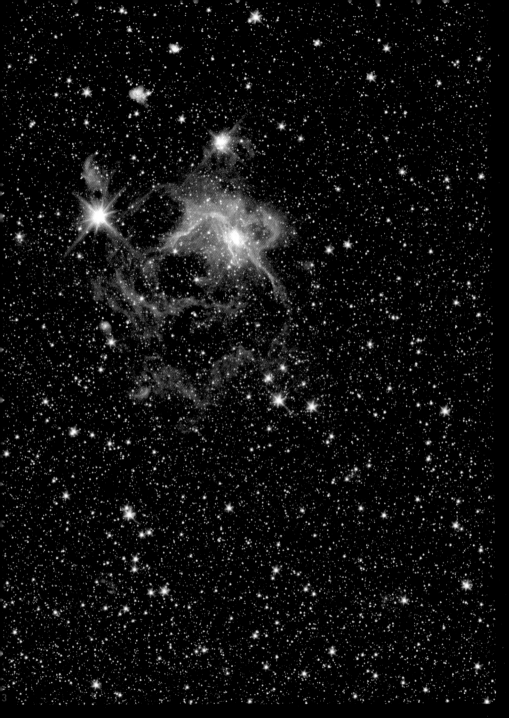

4, 5, 6 RECENT EXPLOSION

In 1992 a tremendous supernova occurred in the constellation of Cygnus [4, 5]. Dubbed Nova Cygni 1992, this event most probably occurred in a binary system. Its smaller white dwarf companion star had so much gas dumped onto its surface from the larger neighbour that conditions became ripe for nuclear fusion. Nova Cygni 1992 was the brightest nova in recent history. This surrounding glow of gas [6] wasn't even visible a few years ago.

10,400 light years

1 GALACTIC SNAKE

The sinuous dark object at the lower left of the frame is a column of sooty cloud. It looks small here, but it could swallow several solar systems. If the Earth happened to pass through this unusually dense material, the starlight would be blocked out and our night skies would appear almost featurelessly black.

10,800 light years

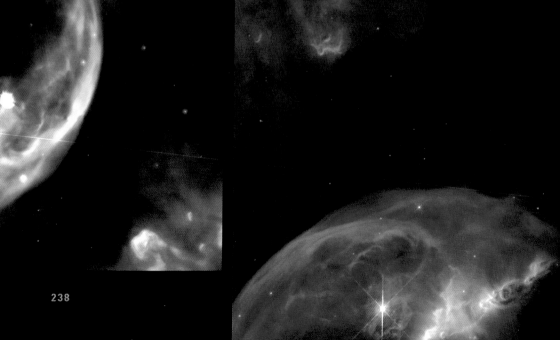

2, 3 BUBBLE NEBULA

NGC 7635, also called the Bubble Nebula, Sharpless 162, or Caldwell 11, is in the constellation Cassiopeia. Although it looks delicate [2] the 10 light-year diameter bubble offers the by-now familiar evidence of a violent star heading towards the end of its life. This more detailed view [3] shows the star, BD+602522, some twenty times as large as the Sun. The bubble seems lopsided because it is ploughing its way through other gas envelopes in the surrounding space.

11,300 light years

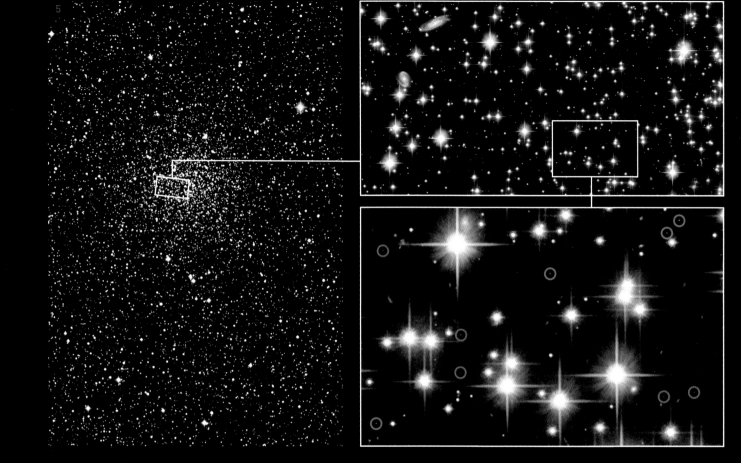

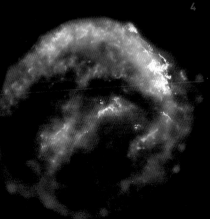

4 KEPLER'S SNR

This supernova in the constellation Cassiopeia was discovered in October 1604, when it was already brighter than all stars in the sky. Several astronomers commented on it, and passed on their observations to Johannes Kepler, after whom it is now named. Initially as bright as Mars, the supernova even surpassed Jupiter's brilliance for a few days. It remained visible until March, 1606.

13,000 light years

5 GLOBULAR PARADOX

The more we observe globular clusters, the less straightforward they seem to be. While studying the dimmest burned-out stars in NGC 6791, the Hubble Space Telescope uncovered a paradox. Stars of widely varying ages exist in an object where all the stars should have formed at the same time. [Left-1] This is a ground-based telescopic view of NGC 6791 in the constellation Lyra.

The green inset box shows the view with Hubble's Advanced Camera for Surveys [top right-2]. These stars are 8 billion years old. Two background galaxies can be seen at upper left [bottom right-3]. A view of a small region reveals faint white dwarfs. The blue circles identify hotter dwarfs that are 4 billion years old, and the red circles show dwarfs that are 6 billion years old.

13,000 light years

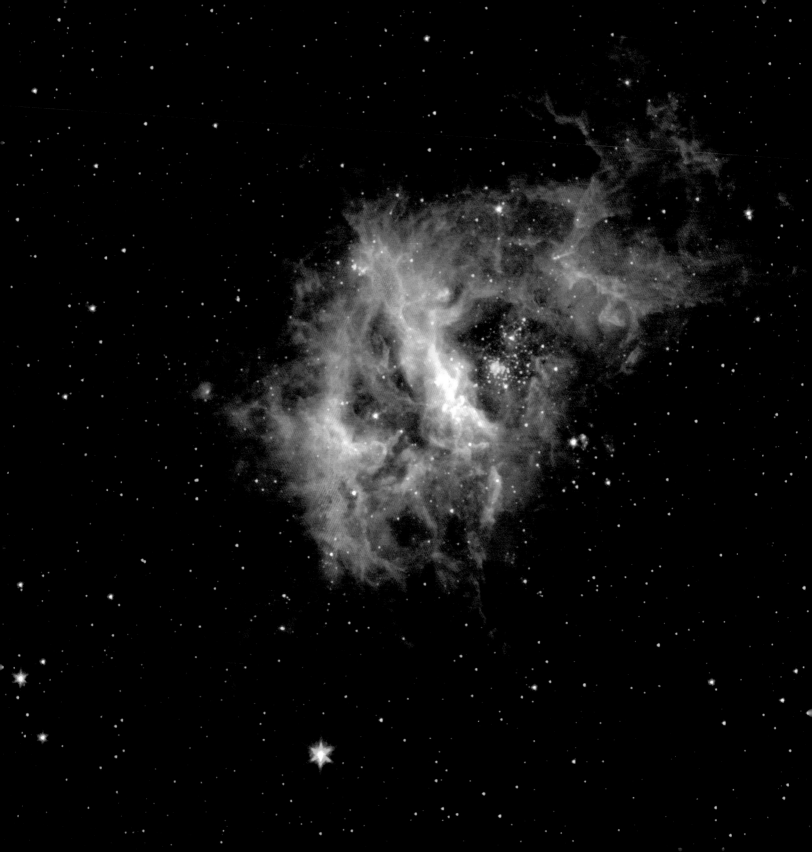

1 Industrious Cloud

This seemingly wispy cloud of gas actually houses more than 2,200 stars. RCW 49 in Centaurus is one of the busiest star factories in our galaxy. A cluster of supergiants have burned away the dusty haze in which they were born. Deeper in the cloud, 300 nascent stars are hiding. Some are surrounded by discs of dust and gas from which planets will form.

13,700 light years

2 Strange lobes

IRAS 19475+3119 in the constellation of Cygnus is a star similar to the Sun that has aged and swollen to become a red giant. The surroundings are rich in dust. The cloud shines by reflecting the brilliant light of the star, and the warmed dust gives off infrared radiation. Jets from the star create strange hollow lobes.

15,000 light years

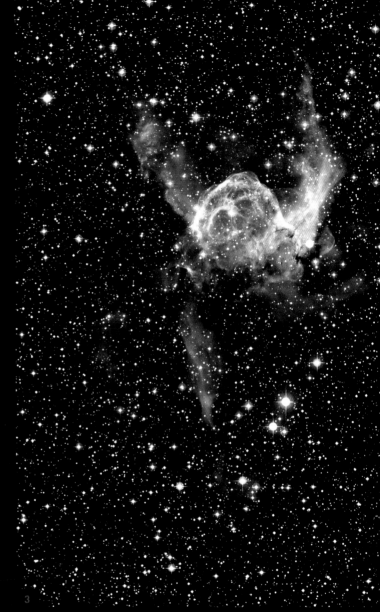

3 About to blow

The emission nebula NGC 2359, or Thor's Helmet, towards the constellation of Canis Major, is shaped by energetic winds from an extremely hot star seen near the centre and classified as a Wolf-Rayet star, a massive blue giant on the verge of supernova. It creates stellar winds with speeds of millions of km per hour. Interactions with a nearby large cloud contribute to the nebula's complex shape and curved bow-shock structures.

15,000 light years

1, 2, 3, 4 **FUTURE CLUSTER**

A favourite observing target of amateur astronomers, Omega Centauri, or NGC 5139 [1,2,3] is a cluster of stars in the constellation Centaurus visible even to the naked eye at low northern latitudes or in the southern hemisphere. It contains 10 million stars, making it ten times more massive than other globular clusters. There may be a black hole at its centre. This multicolour scan [4] captures Omega Centauri's central region and shows the stars moving in random directions, like a swarm of bees. Astronomers charted differences in their positions between 2002 and 2006. From these measurements, they predict future movements [inset]. Each streak represents the motion of a star over the next 600 years. The motion between dots corresponds to 30 years.

16,000 light years

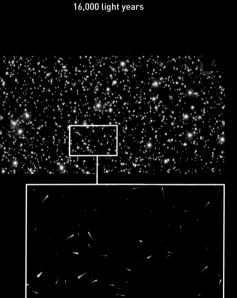

5 PULSAR MAP

Pulsars are spinning neutron stars, collapsed stellar cores left from the final explosions of massive stars. Traditionally identified and studied by observing their regular radio pulsations, two dozen pulsars have now been detected at extreme gamma-ray energies by the Fermi Gamma-ray Space Telescope. This map is aligned with the plane of the Milky Way.

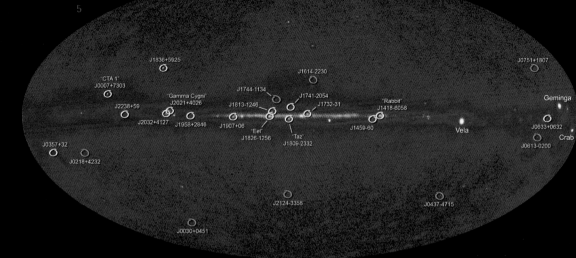

6 SIFTING BY MASS

The globular cluster 7 Tucanae is one of the densest of its type in the southern hemisphere. It contains a million stars, and most of them are arranged according to their mass, governed by a gravitational 'billiard ball' game between stars. Heavier stars slow down and sink to the cluster's core, while lighter stars pick up speed and move across the cluster to its periphery.

16,700 light years

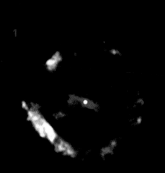

1

1 Uncanny Regularity

When the radio energies of pulsars were first detected in the 1960s, the signals were uncannily like artificial alien pulses because they were so regular, down to fractions of a second. Now we know that pulsars don't pulse. They spin, hurling out wayward beams of energy like cosmic lighthouses. We perceive the pulsing as the beam sweeps across the field of view of our instruments. This is pulsar G11.2-0.3, the remains of a supernova witnessed by Chinese astronomers 1,400 years ago.

16,000 light years

2, 3 IN A DIFFERENT LIGHT

Open star cluster NGC 2467 in the southern constellation of Puppis is really just a coincidental arrangement of stars at different distances than a coherent cluster [2]. Details of dust and clouds in the foreground nebula [3] are revealed by Hubble.

17,000 light years

4, 5 SMALL YET POWERFUL

A small, dense object only 20 km (12 miles) in diameter is responsible for this beautiful x-ray nebula [4]. At the centre of this nebula lies a very young and powerful pulsar, known as PSR B1509-58, a rapidly spinning neutron star that spews energy into the space and creates complex structures, including one that resembles a large cosmic hand [5].

17,000 light years

1 Unsolved Puzzle

Although many planetary nebulae are under observation, astronomers are still puzzled by the 'jets' of energetic material that shoot out of them. The S-shaped 'Garden Sprinker' jets from Henize 3-1475 are among the most perplexing of all.

18,000 light years

2 Stingray Nebula

Just twenty years go, the glow from the shroud of gas surrounding the ailing star was not visible. The nebula, catalogued as Hen-1357, is switching 'on', while the star responsible for its glow is switching 'off', collapsing under its own gravity as the internal pressure of its exhausted nuclear fusion weakens.

19,000 light years

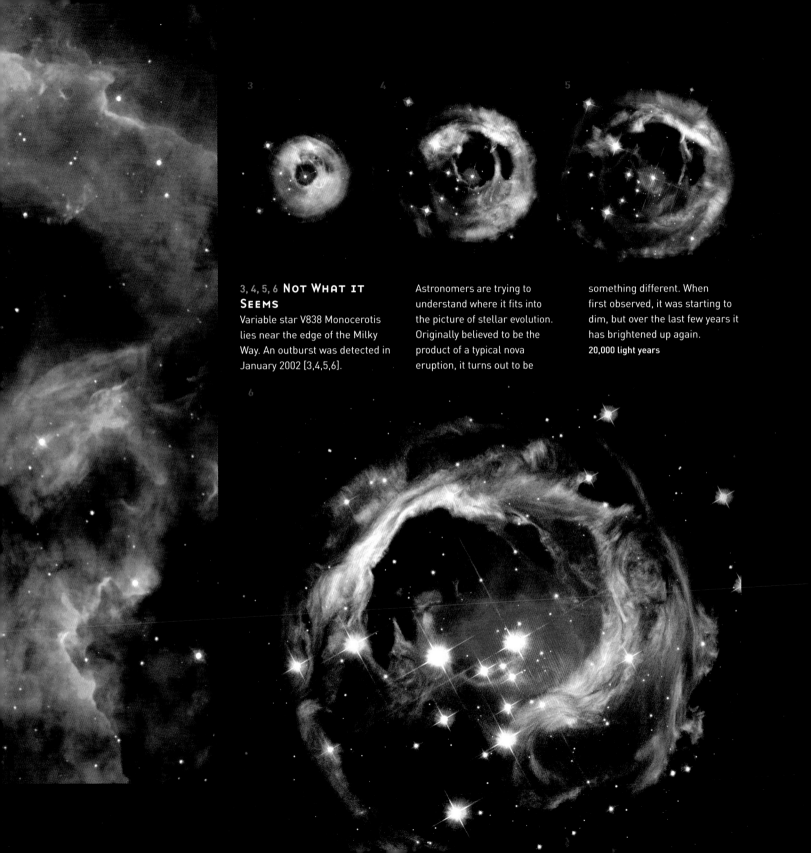

3, 4, 5, 6 NOT WHAT IT SEEMS

Variable star V838 Monocerotis lies near the edge of the Milky Way. An outburst was detected in January 2002 [3,4,5,6].

Astronomers are trying to understand where it fits into the picture of stellar evolution. Originally believed to be the product of a typical nova eruption, it turns out to be

something different. When first observed, it was starting to dim, but over the last few years it has brightened up again.

20,000 light years

1 Most recent supernova

The expanding remains of a supernova explosion within the Milky Way are shown in this composite image of remnant G1.9+0.3. NASA's Chandra X-ray Observatory image obtained in early 2007 is shown in orange [1] and the radio image from NRAO's Very Large Array (VLA) from 1985 is in blue [inset]. The difference in size between the two images gives clear evidence for expansion, allowing the time elapsed since the explosion to be estimated. The initial burst of light reached us 140 years ago. This makes G1.9+0.3 the most recent supernova, as measured in Earth's time frame.

25,000 light years

3 Superstar

Looking into a region of space just 100 light years from the core of the Milky Way, we encounter an extreme environment that gives rise to extreme stars. The young Quintuplet Cluster includes the Pistol Star, a blue hypergiant and one of the most luminous stars in the entire galaxy. Resolved in as much detail as our instruments can manage, the Pistol Star itself hurls out as much energy in a single second as our Sun does in a year.

25,000 light years

2 The Heart of Cluster

This magnificent image from the Hubble Telescope looks into the depths of the Hercules globular cluster, where 100,000 stars are packed into a relatively tight sphere just 150 light years across.

25,000 light years

4 AT THE LIMITS OF THE GALAXY

As we head towards the extreme limits of the Milky Way, ancient globular star clusters roam the perimeter. These spherical groupings of several hundred thousand stars are older than the stars of the main galactic disk.

In fact, measurements of globular cluster ages constrain the age of the Universe, because it must be older than the stars in the clusters, or else something is wrong with our calculations. Star cluster NGC 6934 is certainly 10 billion years old.

50,000 light years

4

OUR GALAXY'S PLACE IN THE UNIVERSE

The Milky Way Galaxy is part of a gathering of between 30 and 40 galaxies known as the Local Group.

If that numbering seems a little vague, this is because some of the galaxies are distinct and discrete, while others are just loose assemblies of stars whose gravitational destinies are shaped by the Milky Way's more substantial influence. Prominent members of the Local Group include M31, other wise known as the Andromeda Galaxy. Two nearer and somewhat untidy companions to the Milky Way are the Large Magellanic Cloud (LMC) and the Small Magellanic Cloud (SMC). There are also several small, shapeless 'irregular' galaxies, and some dwarf elliptical and dwarf ball-shaped galaxies hovering in attendance.

The Large and the Small Magellanic Clouds have been known to astronomers since before the invention of the telescope. They are named after the Portuguese seafaring navigator Ferdinand Magellan, who documented them while circumnavigating the globe between 1519 and 1522. The Large Magellanic Cloud is a complex galaxy in its own right, home to frequent supernovae and numerous star-forming regions, including the famous Tarantula Nebula. As its name suggests, the Small Magellanic Cloud is smaller, and also more distant from us.

The Local Group occupies a volume of space nearly 10 million light years in diameter, with its centre somewhere between the Milky Way and M31. Membership is not certain for all these galaxies, and there are other possible candidates, too. Certainly the Milky Way and M31 are the most massive, and therefore the gravitationally dominant members. The third largest galaxy in the Group, M33 Triangulum, may be a companion hovering around Andromeda, and itself probably has the dwarf galaxy LGS 3 as a satellite.

THE MILKY WAY *Because we live inside our local galaxy, we never get to see it from afar. However, we do know enough for computers to simulate what it would probably look like if seen from alien astronomers in another nearby galaxy.*

1 Multi-spectral Panorama

Marking the International Year of Astronomy in 2009, Hubble, Spitzer and Chandra orbiting instruments collaborated to make this image of the central region of our Milky Way. Infrared and x-ray cameras see through the obscuring dust and reveal intense activity near the galactic core. Yellow represents the near-infrared observations of Hubble, showing regions of star birth. Spitzer's red-hued contribution shows radiation and winds from stars creating glowing dust clouds. Blue and violet x-ray observations from Chandra show gas heated to millions of degrees by stellar explosions, and by outflows from the supermassive black hole in the galaxy's centre.

2 The Galactic Disc

We are surrounded by stars, but the pattern of the galaxy as a whole is disc-shaped and much thinner from top to bottom than edge to edge. This panoramic mosaic focuses on penetrative starlight frequencies while eliminating the blocking effects of dust. Our view is edge-on towards the galactic centre.

3 Stars Like Dust

It is almost impossible to grasp the sheer number of stars that exist, just in our galaxy, even before we try to count up stars in the universe as a whole. This view looking toward the galaxy's centre, and equivalent to an area of sky that could be blocked by a fist held at arm's length shows 10 million stars and a great swathe of galactic dust. At infrared frequencies, the swirls and concentrations of dust are resolved in yet more detail.

4, 5, 6, 7, 8 The Monster at the Core

Like all galaxies, the Milky Way has a colossal black hole at its centre. It is known as Sagittarius A* (pronounced 'A-Star'). It probably contains a mass equivalent to two and a half million Suns. In this x-ray image [8] the region dominated by Sagittarius A* is a bright triangle at upper left. The clouds of gas are reflecting x-rays emitted by matter plunging into the black hole. Closing in [5, 6, 7], we can monitor changes in the clouds. These frames, taken a year apart, show changes that look subtle to our eyes, but which represent huge variations in the story over incredibly short spans of time. At optical wavelengths, we see stars in the same region [4], but no evidence of the black hole. Yet those stars are moving much more swiftly than stars in more outlying regions, as the black hole tugs at them.

1 AGITATED ENERGIES

Radio telescopes can 'see' vast whirls of gas being drawn towards the Sagittarius A* black hole at the heart of the Milky Way. The red blob is not the hole itself, but the shell of extremely agitated material surrounding it. Black holes might seem appalling, yet their gravitational influence may be one of the drivers of galaxy formation and the gathering of materials for new stars.

2 AT THE CORE

Star populations are denser in the galaxy's heart than at its perimeter. In this x-ray scan, regions of gas are being heated up by dense clusters of stars. The closer we see towards the galaxy's heart, the more violent the story gets.

3 FEEDING THE MILKY WAY

At the eleven o'clock and five o'clock positions of this galactic radio scan, gas clouds are being drawn towards the main disc of the Milky Way, seen edge-on in white. Our galaxy is gaining mass by feeding on material from beyond its immediate borders.

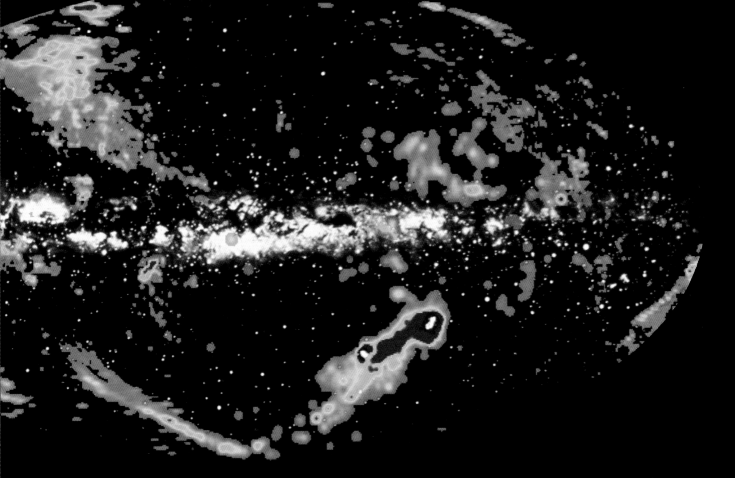

4 Nothing Too Small

The larger image here, covering a portion of the sky about 55 degrees wide, was obtained by the Planck High Frequency Instrument. The dark horizontal band corresponds to the plane of our galaxy, seen in cross-section from our vantage point. The colours represent the intensity of heat radiation by dust. The tighter detail at left was captured by the

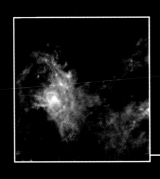

Herschel Space Observatory. Filamentary structures in the galaxy are apparent at large scales and small.

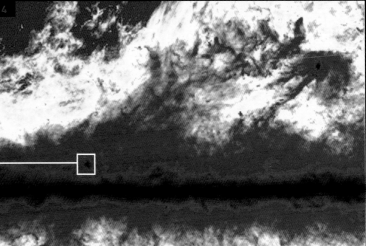

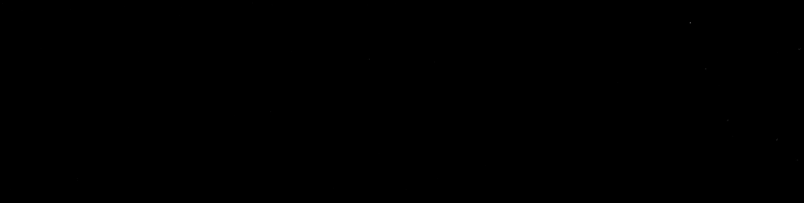

The magnificent Andromeda Galaxy, or M31, is more than two million light years distant, but this is the 'island universe' that gives us our best view of a galaxy somewhat similar to our own.

From our perspective we can see the spiral arms, the dust and gas, even the individual stars within Andromeda's disc. At some 260,000 light years wide, it is twice the size of the Milky Way, although it may be less densely packed with stars. The gap between Andromeda and us is closing at 500,000 km (310,000 miles) per hour. While most galaxies are rushing away from the Milky Way as the universe expands, Andromeda is the only large spiral galaxy moving in our direction.

Even though the universe is expanding, and most populations of galaxies are rushing away from each other, their distribution is uneven. They tend to exist in clusters, bound together gravitationally and constantly influencing one another at localized scales. At least, when we observe a particularly beautiful and neat spiral galaxy such as Andromeda, we can be fairly sure that it has led as placid a life as any galaxy can. However, it looks as if that peaceful circumstance may be destined to change, for Andromeda and the Milky Way alike.

The two galaxies are a gravitationally linked pair. Both galaxies formed alongside each other soon after the Big Bang, then drifted apart with the overall expansion of the universe. But their gravitational attraction for one another has proved too strong, and they are now falling back towards each other. In three billion years from now, they could merge.

Incredibly, as one galaxy passes through another, hardly any of the stars among the mingled billions will collide, because the space between individual stars is so vast. Most of the disruption will occur on the vast scales of gravitational influence. As the two galaxies draw closer, their mutual pull will reshape huge clouds of gas and dust, with the larger galaxy inevitably pulling material away from the smaller victim. Individual stars may not collide that often, but if the black holes lurking in each galaxy's heart come too close, their encounter could have dramatic consequences. Instead of merging together, the forces involved may be so extreme that one black hole will be kicked away at a tremendous velocity. That could cause havoc amid the surrounding star systems.

Perhaps luckily for any galactic inhabitants, galactic collisions are long drawn-out events that take hundreds of millions of years to play out. Even so, the Sun will still be burning when Andromeda and the Milky Way finally meet, and maybe life of some sort will still exist on Earth. As Andromeda approaches, it will grow in size, until the night sky is filled by a giant spiral galaxy. When the two finally intersect, the familiar band of the Milky Way will be joined by a second intersecting arch of stars, and this new pattern will only last for around 100 million years.

After that, the distribution of stars will become muddled and confusing, at least from the perspective of our astronomers, if any still exist by then. There is a slight chance that our Sun, and its accompanying planets, will be ejected into the deeper reaches of space. We can only guess if any humans, or strange descendants of our species, will be around to record these epic events. Whatever happens, it will be perfectly typical behaviour for interacting galaxies.

M31 ANDROMEDA GALAXY *This beautiful whirlpool of dust and stars is the next-nearest major galaxy to the Milky Way. The diffuse light is generated by hundreds of billions of stars. Several apparently bright objects surrounding M31 are foreground stars in our own galaxy.*

ISLAND UNIVERSES

A galaxy is any large and distinct collection of stars dust and gas whose shape comes mainly from the collective gravitational influence of all the material within it.

In terms of their generalized physical appearance in optical light, galaxies are classified in three broad categories, based on the character of the central bulge, or 'nucleus', where star populations are densest, and the surrounding flat disk. Almost all galaxies are rotating like giant wheels. This is a product of residual momentum from the gravitational accretion processes that formed the galaxy. The slow, stately rotation typically draws out long strands or 'spiral arms' of stars at the outermost edges.

Irregular galaxies are usually small and untidy. Either they are being pulled out of shape by the influence of other galaxies, or their gravitational fields are not sufficiently powerful to shape them neatly. Elliptical galaxies are basically all nucleus, with little or no surrounding disk. They can range from spherical to elongated football-like shapes. Spiral galaxies are conspicuous for their spiral-shaped arms emanating from the nucleus and gradually winding outwards to the edge of the disk. There are usually opposing sets of arms arranged symmetrically around the centre. So-called 'barred spiral' galaxies have arms extending from the ends of an approximately bar-shaped nucleus.

THE MILKY WAY *Huge star clouds dominate this view of the sky towards the centre of our galaxy, but between them lies something else – deep veins of absorbent dust and excited gas, and the glowing pink rosettes of star-forming nebulae. Stars are mere ephemera – gone in a few billion years at most.*

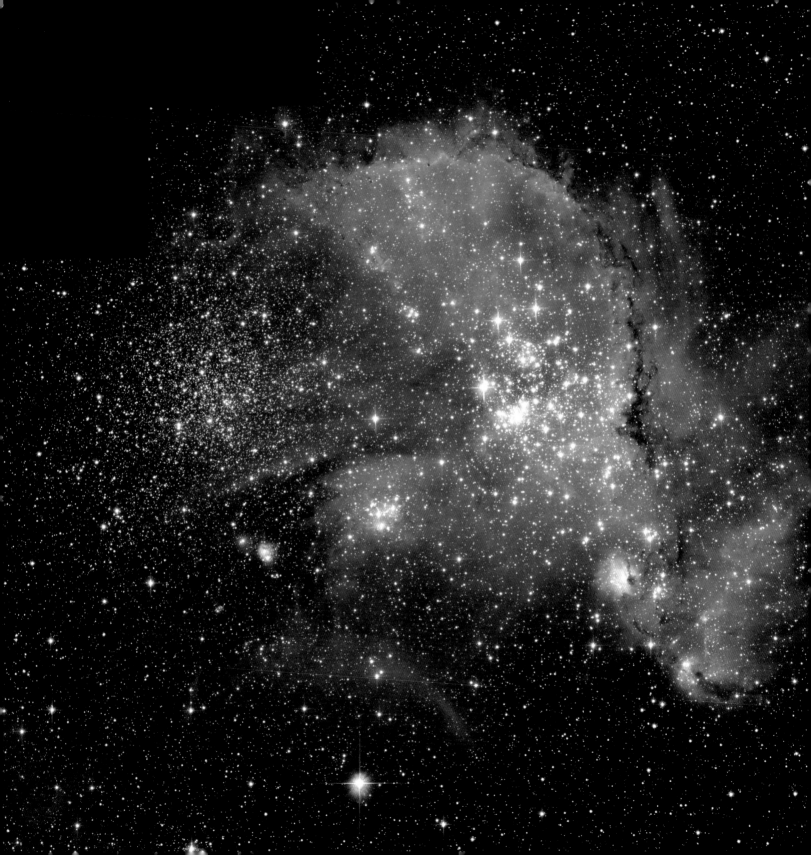

1

1, 2 UNTIDY MARGINS

In our mind's eye, the Milky Way is a discreet disc of 200 billion stars with a neat perimeter of attractive spiral arms. The truth is not quite so tidy. Nearby irregular 'dwarf galaxies' fall under its gravitational influence. The Large Magellanic Cloud [1] and its apparent relative, the Small Magellanic Cloud [2], are conspicuous in the southern hemisphere. They look like partially separated pieces of the Milky Way, and are destined to be consumed by it.

160,000 light years

2

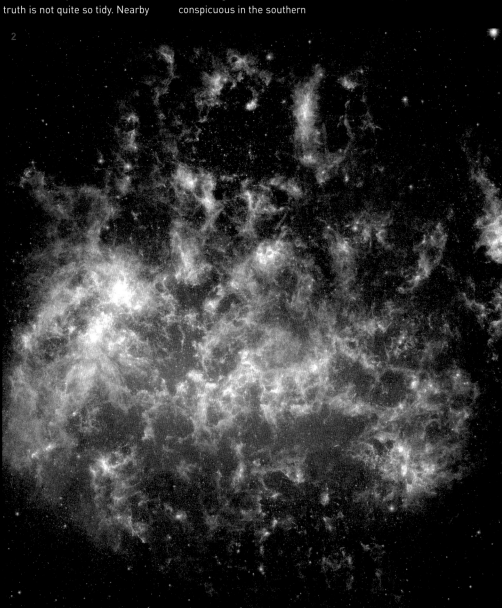

1 MAGELLAN'S HORSE

It may look like a grazing seahorse, but the dark object toward the right is a pillar of smoky dust about 20 light years long, within the Large Magellanic Cloud, and very near the Tarantula Nebula.

157,000 light years

2, 3 RAGGED WRECKAGE

The N49 supernova remnant is the brightest of its kind in the Large Magellanic Cloud. [2] Most supernova remnants adopt a roughly ball-shaped structure. This one is unusual because its filaments appear unusually untidy, no matter what wavelengths we use to observe them [3].

160,000 light years

4 FERTILE FIELDS

This detail of the Large Magellanic Cloud spans about 100 light years, and shows the fertile LH 95 star-forming region with stars ranging from blue giants to nascent 'protostars'.

160,000 light years

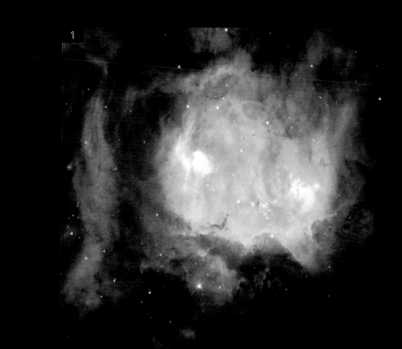

1

2

1 A GHOST IN THE CLOUD

Nebula NGC 2080, otherwise known as the Ghost Head Nebula, is yet another star-forming region belonging to the Large Magellanic Cloud. Hydrogen and oxygen clouds are being superheated by newborn stars within.

160,000 light years

2 A SMALLER BANG

Our usual assumption is that only very large stars die in massive supernovae. Modest-sized stars can end with a bang, too, so long as they draw in additional mass first from a hapless companion star that strays too close. This is the shell of DEM L71, another of many fascinating objects in the Large Magellanic Cloud.

160,000 light years

3, 4 Bright Spider

If this burst of star formation in the NGC 2070 Tarantula Nebula (also known as 30 Doradus) were as close to the Earth as the great Orion Nebula, the resultant glowing gases would cast shadows like bright moonlight at night. [3] The nebula, safely located in the Large Magellanic Cloud, is home to some of the hottest and largest stars ever encountered [4].

160,000 light years

1 INTO THE Tarantula's Jaws

Another view inside the monstrous Tarantula Nebula shows huge bubbles of energetic gas, long filaments of dark dust, and unusually massive knot of stars so dense that it was once thought to be a single star.

160,000 light years

2, 3 DISSECTING THE Tarantula

These two images, taken in visible and infrared light by the Hubble Telescope's Wide Field Camera, shows a massive star cluster R136 in the Tarantula Nebula. In ultraviolet, visible, and red light, the most prominent stars look like icy blue diamonds [2]. The green in the nebula is from the glow of oxygen and the red is from fluorescing hydrogen. At infrared wavelengths [3] Hubble sees through the dust, revealing stars that are usually hidden.

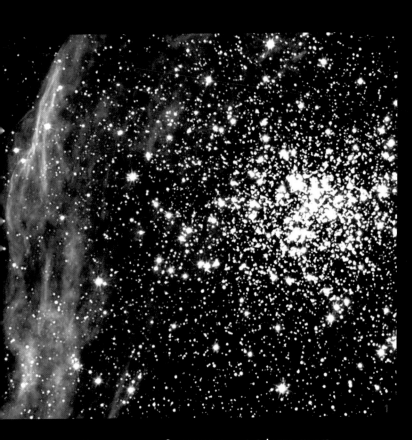

1 MAGELLANIC CLUSTERS

The double cluster NGC 1850 in the Large Magellanic Cloud has two relatively young components. The main, globular-like cluster is in the centre. The smaller cluster, below and to the right, is composed of extremely hot, blue stars and fainter red stars. The main cluster is about 50 million years old, and the smaller one, four million years old.

168,000 light years

2, 3 LATE NEWS

Supernova SN 1987A in the outskirts of the Tarantula Nebula in the Large Magellanic Cloud [2]. A burst of light from the explosion reached Earth in February, 1987, and its brightness peaked in May of that year. Hubble's more detailed view [3] shows dozens of bright spots. Shock waves of material unleashed by the blast are slamming into the inner layers of an earlier debris ring, causing them to glow. The phrase 'are slamming' has to be considered in the context of time as well as space. The events shown here occurred 168,000 years ago.

168,000 light years

3

4 CARVED CAVERN

Fierce radiation from the blue
giant at centre has scorched a
void, some 35 light years across,
in nebula N44F, located in the
northern outskirts of the N44
complex of nebulae in the
Large Magellanic Cloud.
170,000 light years

4

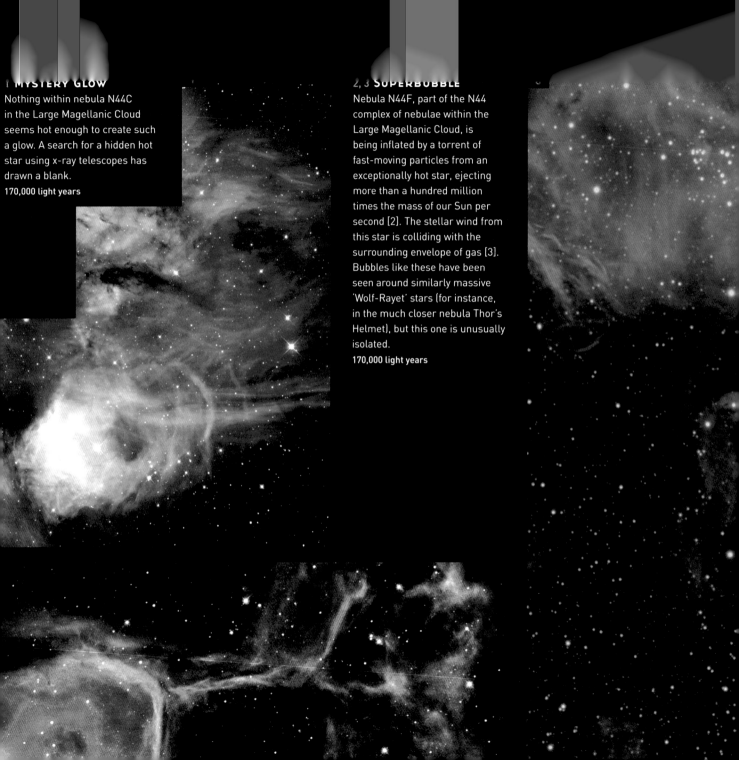

1 MYSTERY GLOW
Nothing within nebula N44C
in the Large Magellanic Cloud
seems hot enough to create such
a glow. A search for a hidden hot
star using x-ray telescopes has
drawn a blank.
170,000 light years

2, 3 SUPERBUBBLE
Nebula N44F, part of the N44
complex of nebulae within the
Large Magellanic Cloud, is
being inflated by a torrent of
fast-moving particles from an
exceptionally hot star, ejecting
more than a hundred million
times the mass of our Sun per
second [2]. The stellar wind from
this star is colliding with the
surrounding envelope of gas [3].
Bubbles like these have been
seen around similarly massive
'Wolf-Rayet' stars (for instance,
in the much closer nebula Thor's
Helmet), but this one is unusually
isolated.
170,000 light years

1, 2, 3, 4, 5 Analysis of a cataclysm

The shape of supernova remnant E0102-72 provides astronomers with clues about how tremendous explosions disperse chemical elements into space [1]. Radio waves, shown in red [2] trace high-energy electrons spiralling around magnetic fields in the shock wave expanding from the detonated star, while optical light, in green, traces cool gas, including oxygen. Finally the x-ray analysis [3] shows gas heated by an inward-moving shock wave rebounding from its collision with the slower-moving outer layers. Similar investigation of another remnant, N63A, tells a similar story [4,5].

190,000 light years

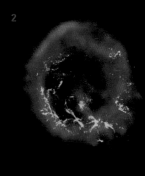

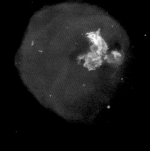

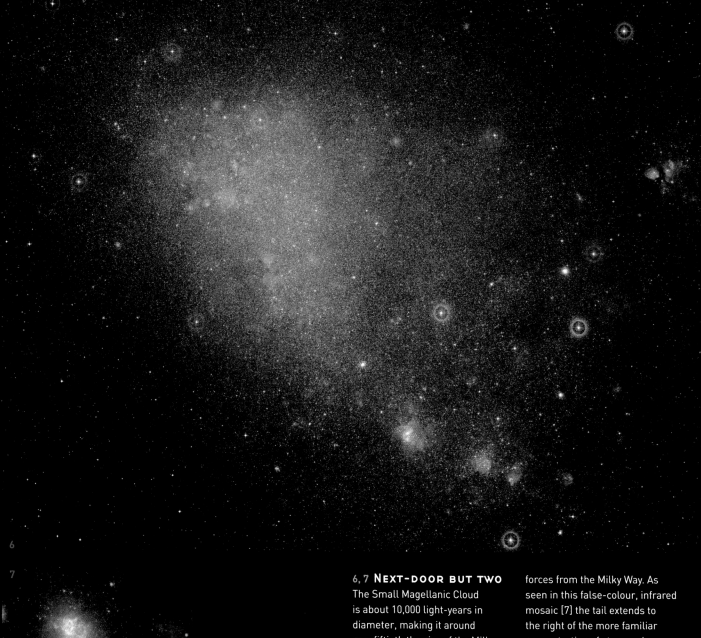

6

7

6, 7 NEXT-DOOR BUT TWO
The Small Magellanic Cloud is about 10,000 light-years in diameter, making it around one-fiftieth the size of the Milky Way [6]. It is the third nearest of the neighbouring Local Group galaxies, and is visible to the naked eye in the southern constellation Tucana. It also has a tail of gas, dust and newly formed stars, a product of tidal forces from the Milky Way. As seen in this false-colour, infrared mosaic [7] the tail extends to the right of the more familiar concentration of stars and gas. Two clusters of newly formed stars, warming their surrounding dust clouds, are seen in the tail as red spots.
200,000 light years

Near and Far Wonders

Near the outskirts of the Small Magellanic Cloud lies the young star cluster NGC 602, surrounded by interstellar gas and dust. NGC 602's massive young stars have eroded the dusty material and triggered a progression of star formation progressing outwards from the cluster's centre. This image spans about 200 light-years. A wide assortment of background galaxies, millions of light years further distant, are also visible.

200,000 light years

THE DRAMATIC LIVES OF GALAXIES

Galaxies can lead violent lives, pulling each other apart, smashing into one another, or lighting up vast regions of space with immense bursts of electromagnetic energy.

Seyfert galaxies, named after the astronomer Carl Seyfert who first discovered them in the 1940s, account for just one or two in every hundred spiral galaxies. They may be rare, but they certainly announce their presence. Their cores (nuclei) change brightness with startling rapidity at intervals of just a few weeks. Whatever causes those fluctuations must be very small, yet at the same time, immensely powerful. The rotational velocities at the centre of Seyfert galaxies are about thirty times greater than we would expect to find in the centres of more peaceful spiral galaxies.

Radio galaxies are, as their name implies, 'loud' sources of radio energy, probably coming from jets of subatomic particles interacting with magnetic fields associated with black holes. Quite apart from their tremendous internal energies, galaxies lead active lives while on their travels, often piling into one another. Colliding or 'interacting' galaxies are found throughout the universe, and these crashes usually trigger fresh bursts of star formation. These mergers may form new and bigger galaxies, even if some hapless stars are hurled away in the chaos, their destiny to drift in near-empty regions of intergalactic space as lonely outcasts. Other galactic interactions are more like hit-and-run car crashes. One galaxy punches through another, tears apart its victim and sails on, bruised but more or less intact.

INTO THE DISTANCE *Although Hubble-V and Hubble-X emission nebulae look similar to the nearby Orion Nebula, they are much brighter and bigger, and located far beyond the margins of our galaxy. They lie within NGC 6822, an irregular galaxy approximately 1.6 million light-years away in the constellation Sagittarius. We are still within the 'Local Group', but now the distance scale is approaching spans of space and time that are no longer easy for the human imagination to grasp (1.6 million light years).*

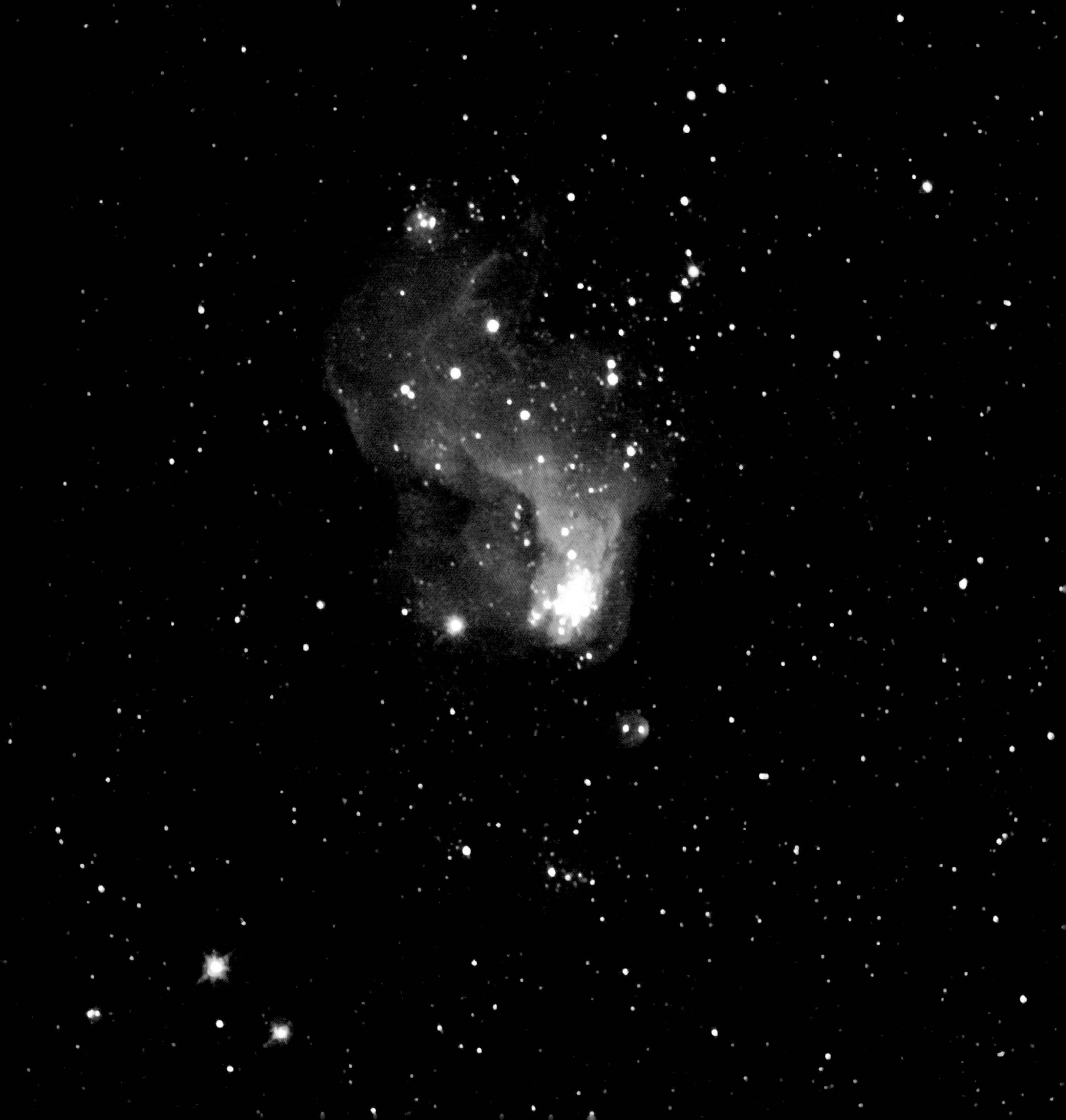

1 What Makes a Galaxy?

The stars that are so visibly obvious account for only a small proportion of the mass that makes a galaxy. Most of the material is gaseous hydrogen, as shown in these 'hydrogen portraits' of nearby galaxies, obtained using radio telescopes that specifically cancel out images of stars.

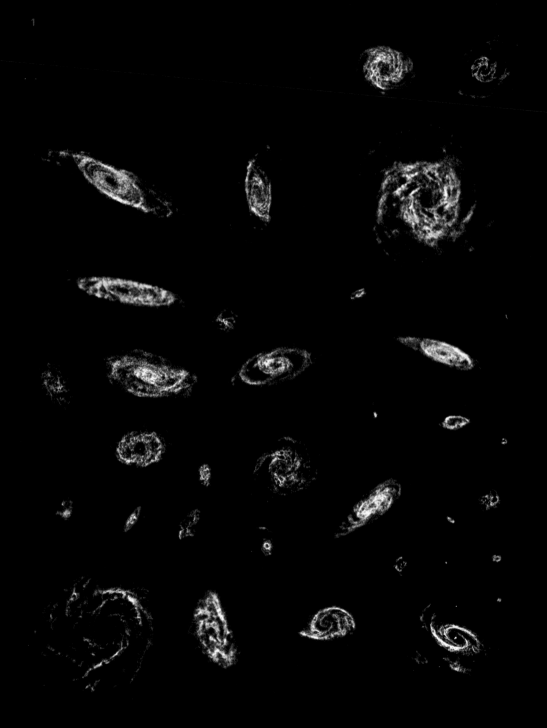

1

2 Largest in Red

This survey shows the thirty largest galaxies as seen in infrared light. Interestingly they do not match the thirty largest optically visible galaxies. One key reason is that the millions of stars whose combined light forms the nebulous shape of a galaxy are brighter in the visible part of the spectrum. Galaxies that appear faint in visible light may seem much smaller in the infrared, or indeed may not be detected at all. Also, in large galaxies, the visible light can be strongly dominated by a relative handful of the most massive stars. In the infrared, this effect is greatly reduced and the observed light is a better tracer of the actual distribution of star mass. When seen from the edge, the increased thickness of stars against the background of deeper space makes even the faintest galaxies more visible. Therefore, edge-on galaxies are more likely to be detected and will tend to appear larger.

The Ultimate Galaxy

Most galaxies are too far away to be appreciated properly, even in our best instruments. Our own Milky Way surrounds us, confusing our understanding of its shape. Most of the Local Group galaxies are, to be blunt, untidy affairs, pulled out of shape by their proximity to the Milky Way. However, one member of that group is sufficiently large that it controls its own gravitational destiny, at least for the next few billion years to come. In the Andromeda Galaxy we see something very like what our Milky way would like from the outside, except that Andromeda is twice as large.

2.5 million light years

1, 2, 3, 4, Influences, Great and Greater

Pictured here in these 45 images are 15 recent supernovae, each shown in three wavelengths: optical, ultraviolet and x-ray [1]. A series of 'mug shots' capture the host galaxy surrounding each violent star explosion. These localised dramas influence star formation in the galaxies. Over immense spans of time, gravitational rearrangements at local scales subtly effect the shaping of vast galactic populations as they cluster under the influence of their mutual gravitational attraction, visualised in these simulations [2,3,4]. The lives and deaths of individual stars might seem insignificant at cosmic scales, but no event is too small to count.

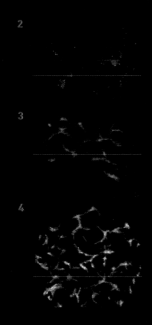

2

3

4

5 X Marks the Spot

Hubble-X is a glowing gas cloud in one of the most active star-forming regions within galaxy NGC 6822, one of the Milky Way's closest neighbours, locatable in the constellation Sagittarius. This hotbed of star birth is similar to the fertile regions in the Orion Nebula, but on an even grander scale.

1.6 million light years

5

1, 2, 3, 4 Analysing Andromeda

The fraction of light energy that human eyes can detect reveals relatively little of the matter and energy shaping celestial objects. In this composite image of Andromeda [1, 2, 3] infrared heat energies are shown in red and orange, and intense x-ray energies in blue and white. A separate ultraviolet analysis [4] picks out the youngest and hottest stars.

5 Collisions in Andromeda

Around one third of the 300,000 stars in Andromeda's outer halo are only 6–8 billion years old. The presence of these comparatively young stars, just half the age of most of their more inwardly located neighbours, suggests that a collision with a smaller and long-since dissipated galaxy might have created a wave of star formation.

6 Unimaginable Vastness

Beautiful Andromeda is just one galaxy among billions. A thin strip of the night sky, near the familiar constellation of the Plough, or Big Dipper, investigated at different telescope magnifications tells us that the deeper and closer we look, the more we find distant specks of light resolving themselves as entire galaxies, island universes, each with hundreds of billions of stars.

7 Black holes

The Chandra X-ray Observatory has made regular observations of the Andromeda Galaxy's core, watching for variations in energy levels over more than a decade. Astronomers have made an unprecedented study of Andromeda's supermassive black hole, the nearest outside the Milky Way. It was in an unusually calm state until January 2006, when it suddenly became a hundred times brighter, suggesting an outburst of x-rays. This was the first time such an event had been seen from anywhere within the nearby universe. The cause of the outburst is unknown, but it could have been connected with immensely powerful magnetic field lines suddenly converging.

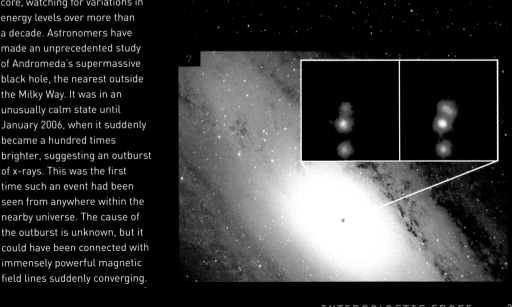

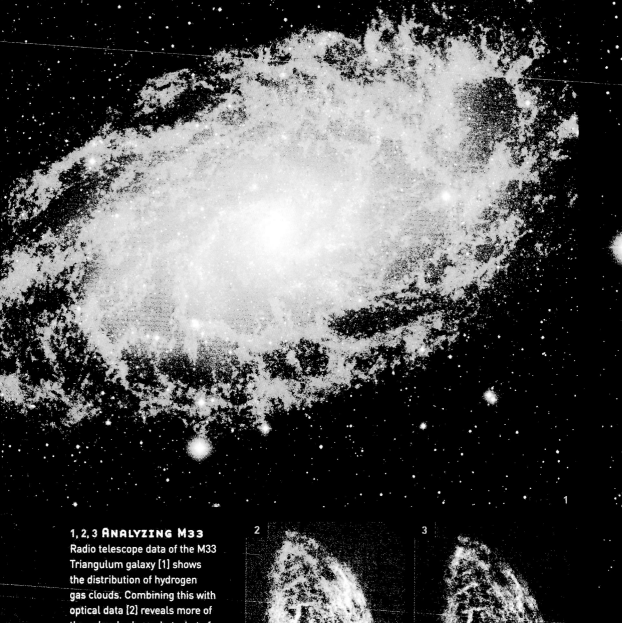

1, 2, 3 ANALYZING M33

Radio telescope data of the M33 Triangulum galaxy [1] shows the distribution of hydrogen gas clouds. Combining this with optical data [2] reveals more of the galaxy's shape, but what of its dynamics? Doppler shifts in the radio frequencies betray the galaxy's rotation [3]. The blue side is rotating towards us.

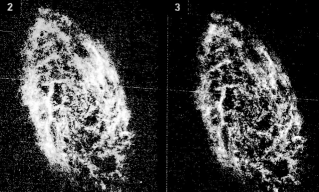

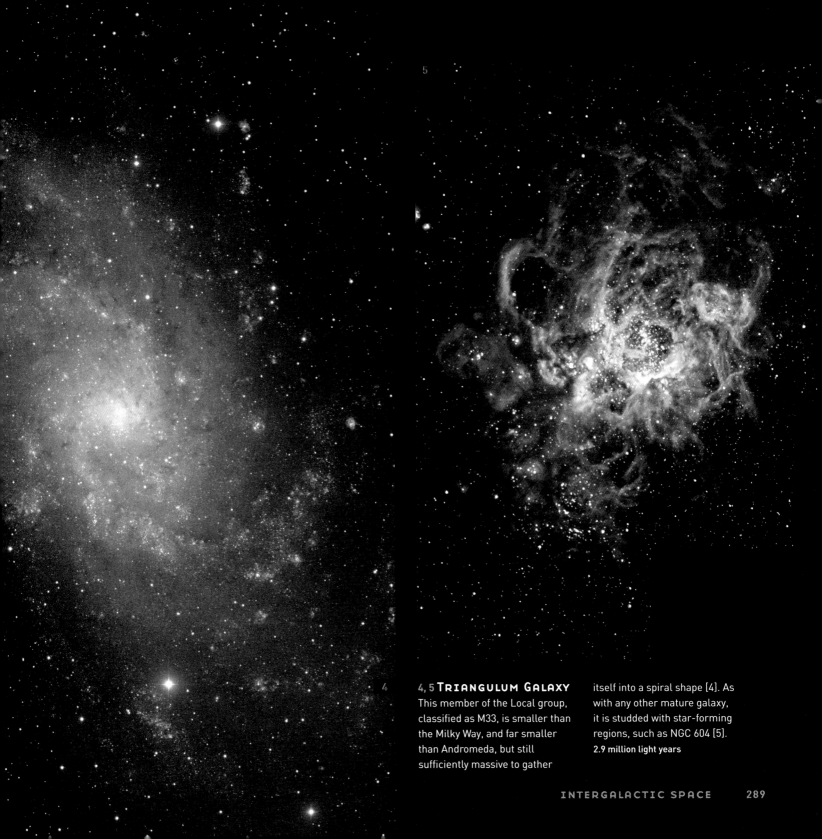

4, 5 Triangulum Galaxy

This member of the Local group, classified as M33, is smaller than the Milky Way, and far smaller than Andromeda, but still sufficiently massive to gather itself into a spiral shape [4]. As with any other mature galaxy, it is studded with star-forming regions, such as NGC 604 [5].

2.9 million light years

1 DIVERSE STRUCTURES

These images from the Hubble Space Telescope are close-ups of four galaxies, each with very different masses and sizes. Although they are separated by many light years, they are shown at relative scales. In the composite image at the top, NGC 253, in the Sculptor Group of galaxies, is ablaze with the light from thousands of young, blue stars. Within NGC 300, second from top, similarly young stars are concentrated in spiral arms sweeping diagonally through the image. The dark clumps of material scattered around the core of NGC 3077, the small, dense galaxy at bottom, left, are pieces of wreckage from interactions with that galaxy's larger neighbours. The image at bottom right, shows a swarm of young, blue stars in the diffuse irregular dwarf galaxy NGC 4163

6.5 million to 13 million light years

2 NGC 300 SPIRAL

This galaxy was discovered by Scottish astronomer James Dunlop, who spotted it from Australia in the early 19th century, although the scale and significance of galaxies was not properly appreciated until the 20th century. It is located in the constellation Sculptor, which contains a few bright stars but is made up of a collection of galaxies that form the Sculptor Group.

7 million light years

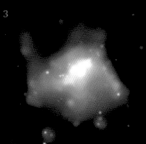

3, 4 IRREGULAR GALAXY

Sporadic bursts of star formation
are pushing this nearby irregular
galaxy out of shape, not that it
had much by way of elegant
structure in the first place [4].
Galaxies tend to settle into
neater shapes under the
influence of their greater
collective gravitational fields.
This x-ray image from the
Chandra Observatory [3] shows
large hot bubbles extending
above and below a disk of gas
along the equator of the galaxy.
Chandra detected oxygen, neon,
magnesium, and silicon in the
bubbles and the disk.

7 million light years

5 INVISIBLE NO MORE

Maffei 2, named after the
astronomer who found it, is
almost invisible to optical
telescopes. Foreground dust
clouds in the Milky Way block
about almost all its visible light.
The emergence of infrared
astronomy enabled this, and
many other galaxies, to be picked
out from behind the intervening
veils of dust.

10 million light years

1, 2, 3 DUSTY LANE

The galaxy Centaurus A, or NGC 5128 in the constellation of the Centaur, is one of the closest and most spectacularly active of its kind [1]. Its radio signature, as imaged by the Very Large Array radio telescope at Socorro, New Mexico, shows jets of energy extending far into space from its active core [2]. Images at a wider range of frequencies, in broader view and in tight detail [3] show an enormous dust lane running across the centre of Centaurus A. This is the remains of another galaxy that collided with it. The dismembered wreckage is now feeding a powerful black hole.

11 million light years

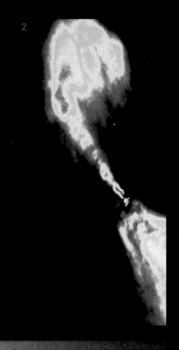

1 MYSTERY CIGAR

An unknown object in the nearby 'Cigar' galaxy M82 is sending out radio emissions that are unlike any similar pulses seen anywhere else in the universe. They do not fit the pattern of emissions from supernovae, which usually get brighter over a few weeks and then fade away over months. The new source has hardly changed in brightness since being detected more than a year ago. Yet it does seem to be moving with an apparent sideways velocity of four times the speed of light.

12 million light years

2, 3 STARS AND DUST

The beautifully arranged M81 galaxy, sometimes known as Bode's Galaxy [2] is in the constellation Ursa Major and is one of the brightest visible in our skies. Its grand spiral arms and sweeping dust lanes are similar in scale to those of our Milky Way. This image is dominated by the light from its core of old yellow and red stars. Switching to an infrared view [3] we see the dust lanes. M81 is a typical 'barred spiral' galaxy. A bar-shaped core of stars gradually merges into the spiral arms.

12 million light years

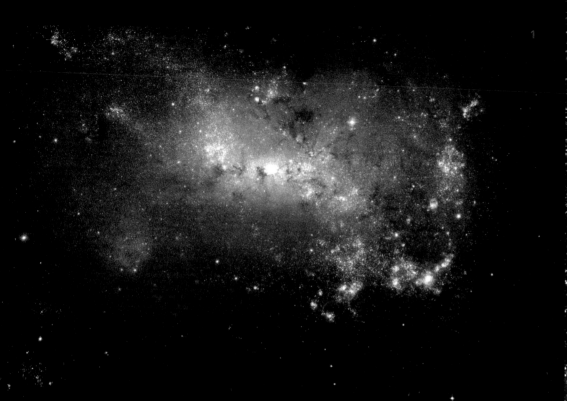

1 BLAZING DWARF GALAXY

NGC 4449 is an irregular galaxy in the constellation Canes Venatici. It is similar in size and structure to the Large Magellanic Cloud. Regions of star birth are not much influenced by the galaxy's ragged shape. This may be a rare example of the young 'primitive' galaxies that we encounter at much greater distances, as our instruments probe the universe's early history.

12.5 million years

2,3 HIDING BEHIND THE CORE

The Circinus galaxy [2] was only discovered as recently as 1975, because it sits almost exactly behind the densely populated core of the Milky Way [3], and so was heavily obscured to earlier observers. It belongs to a class of galaxies called Seyferts, which have compact centres and are believed to contain massive black holes. Seyfert galaxies are themselves part of a larger class of objects called Active Galactic Nuclei, or AGN. The stars alone do not account for their energy output. Black holes are assumed to be a major factor.

13 million light years

1, 2, 3, 4, 5 THE
SCULPTOR'S SCULPTOR

NGC 253 is the brightest member
of the Sculptor Group of galaxies,
grouped around the South
galactic pole [1]. It is also often
referred to as the Sculptor
Galaxy. In this mosaic [5] from
the Wide-field Infrared Survey
Explorer, WISE, the red image at
bottom right [4] shows the
galaxy's active side. Infant stars
are heating up their dusty
cocoons. The green image at
centre right [3] reveals the
galaxy's emerging young stars,
concentrated in the core and
spiral arms. Ultraviolet light
from these hot stars is absorbed
by tiny dust particles left over
from their formation, making the
particles glow with infrared light,
colour-coded green in this view.
The blue image at top right [2]
shows stars of all ages.

13 million light years

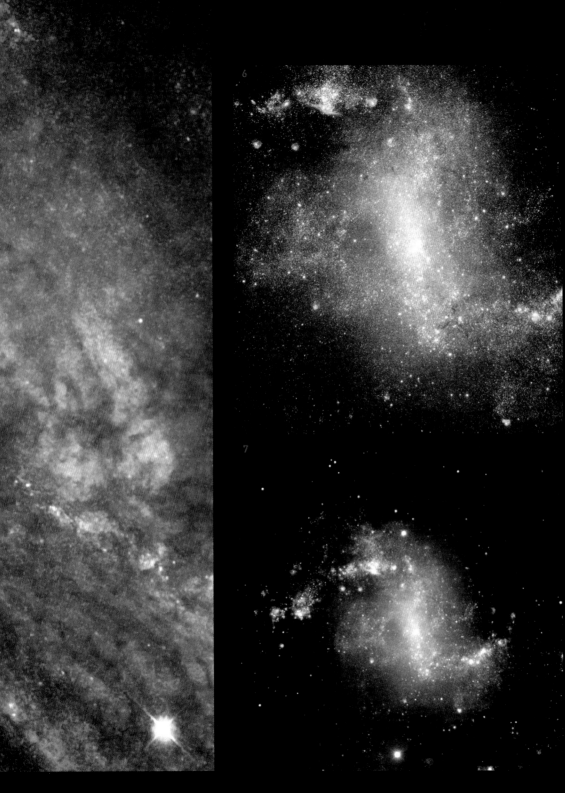

6,7 Off-Centre

Individual stars in the barred spiral galaxy NGC 1313 are resolved in this sharp composite from the Hubble Space Telescope. The inner region of the galaxy is pictured, spanning about 10,000 light-years [6]. A wider view shows that NGC 1313's spiral arms are lopsided [7] and its rotational axis is not at the centre of its bar-shaped core.

14 million light years

1

2

3

1, 2, 3, 4 SOUTHERN PINWHEEL

This galaxy, otherwise known as M83, is one of the closest and brightest spiral galaxies in the sky, visible even through binoculars in the constellation of Hydra [1]. It is a prominent member of a galactic group that includes Centaurus A and NGC 5253. Several bright supernova explosions have been recorded in M83. Its x-ray portrait [2] shows hot gas in the spiral arms, and superheated gas at the core, where point sources of intense energy, probably in the form of neutron stars and black holes, are more densely packed. Hubble's closer inspection of M83's spiral arms [3] identify a range of globular clusters and several dozen supernova remnants. Radio observations [4] map out the signals from hydrogen gas, showing how the spiral arms of M83 extend far deeper into space than their optically visible traces would suggest. The blue whorl in the centre is the visible portion.

15 million light years

4

1 DELICATE BRIDGE

Two galaxies, NGC 5216 and NGC 5218 in the constellation Ursa Major are linked by an usual filament of luminous debris.

17.3 million light years

2 BLACK EYE GALAXY

This galaxy, M64, gets its name in honour of the conspicuous dark structure close to the nucleus. Another oddity is that the interstellar gas in the outer regions rotates in the opposite direction from the gas and stars in the inner regions. This may be a legacy from a collision with another galaxy more than a billion years ago.

17 million light years

3, 4 Neat and Orderly

Spiral galaxy NGC 2841 in Ursa Major [3] has a very organized structure, with stars sprinkled evenly throughout. 'Rivers' of young stars drift from their hot, dense stellar nurseries and disperse to form large, smooth distributions. With a diameter of over 150,000 light-years, it is even larger than the Milky Way. This composite image [4] shows the older stars, while cooler areas of dust and gas are highlighted in red.

19.6 million light years

3

4

1, 2 Warped arms

The uneven structure of this highly active Seyfert galaxy, M106 in Canes Venatici [1] reveals major disturbances. We can only assume that gravitational interaction with other nearby galaxies caused the distortion, which is even more obvious when radio imaging shows additional arms [2] highlighted in blue.

22 million light years

4 EDGE-ON SPIRAL

A Hubble's-eye view of NGC 2683 in the northern constellation of the Cat, with more distant galaxies scattered in the background. Light from a large population of old yellow stars forms the remarkably bright galactic core. Starlight silhouettes the dust lanes along the spiral arms. The occasional patches of glowing blue are from young stars in this galaxy's star-forming regions.

20 million light years

3 CONCENTRIC RINGS

Around the centre of this 'lenticular' galaxy NGC 2787 in Ursa Major, almost concentric arms of dark dust encircle the bright nucleus. Lenticular galaxies are so-called because they are shaped somewhat like glass lenses.

24 million light years

1, 2, 3, 4 BILLIONS UPON BILLIONS

The Pinwheel Galaxy, or NGC 5457 in Ursa Major [1, 4] is nearly twice the diameter of the Milky Way, and host to a thousand billion stars. An ultraviolet study [2] picks out energetic young, hot stars. More faces of M101 are apparent in this composite of views from Spitzer, Hubble, and Chandra [3]. Red shows infrared heat emitted by dust lanes. Yellow is the visible light. Most of this light comes from stars, and they trace the same spiral structure as the dust lanes. In blue are the x-ray sources, including million- degree gas, exploded stars,and material around black holes.

27 million light years

1

2

3

4

1, 2, 3 SOMBRERO GALAXY

As seen from Hubble's vantage point in Earth orbit, galaxy M104 at the southern edge of the Virgo cluster [1] is tilted nearly edge-on. It gets its familiar name because of its resemblance to the broad-rimmed Mexican hat. The Sombrero hosts at least 2,000 globular star clusters, ten times as many as are associated with the Milky Way. Embedded in the bright core of M104 is a smaller disk, is tilted relative to the larger one. In here lurks a black hole with the mass of a billion suns. Resolved at infrared frequencies, the dust makes a prominent ring [2] while x-ray concentrations [3] show the most energetic sources.

28 million light years

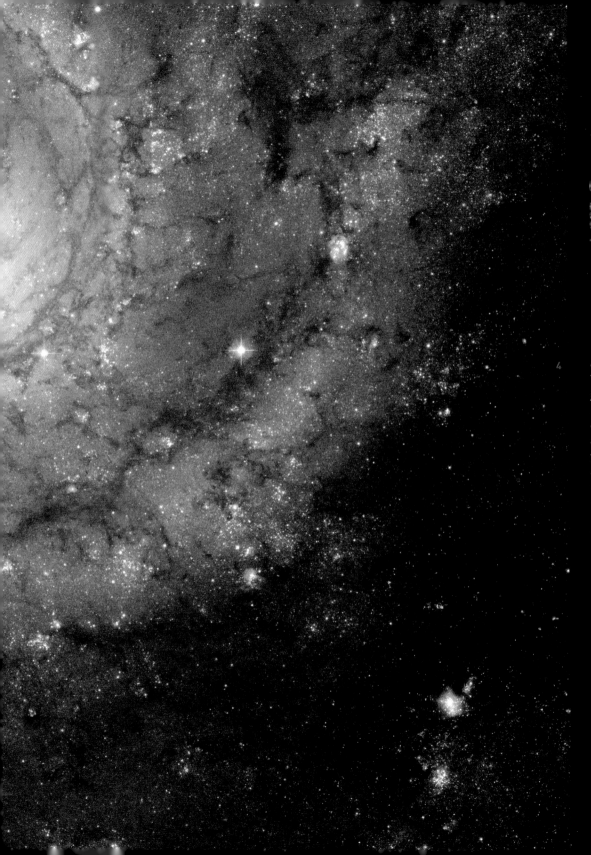

5

4, 5 PERFECT SCORE

The M74 galaxy is situated
just northwest of the star Eta
Piscium, the bright one in the
constellation Pisces. The galaxy's
arms [4] span about 80,000 light
years. Its structure is classified
as a 'Grand Design Spiral'
because it is such a perfect
example of its kind. Multi-
spectral analysis [5] shows
the familiar whirls of dust in
infrared, and the power of hot
stars identified in x-rays. As with
most sizable galaxies, the core
hides a violently powerful heart,
probably in the form of a
supermassive black hole.

32 million light years

4

1 MODERATELY ANCIENT LIGHT

Ultraviolet and visual image comparisons of the barred ring galaxy NGC 1291 in the constellation Eridanus. At this range, we are heading deep into the past as well as merely physical distance. The light that we see here began its journey long before humans first emerged on the Earth. Even so, this is a comparatively 'young' object in a universe that came into being 13.75 billion years ago.

33 million light years

3, 4, 5 WHIRLPOOL GALAXY

The appropriately-named galaxy, more formally known as M51, is one of the most conspicuous, and probably the best-known spiral galaxies in the sky after Andromeda [3]. Seen in near-infrared light [4] with most of the optical starlight removed, the Whirlpool's complex dust structure is revealed. Located in the constellation Canes Venatici, and classed as a Grand Design spiral, M51's shape must be influenced by the mutual gravitational forces between it and its very obvious neighbour, irregular galaxy NGC 5195, seen here in a wider view [5].

37 million light years

2 ROUTINE SPLENDOUR

Spiral galaxy NGC 3521 toward the constellation Leo is approximately 50,000 light years in diameter. It has multiple irregular spiral arms laced with dust and clusters of young, blue stars. In the great scheme of existence, this celestial wonder is just a relatively ordinary patch of the cosmos.

35 million light years

1 SHIFTED REDSHIFT

We are not quite sure how far away NGC 4216 really is. This edge-on spiral galaxy is located not far from the centre of the Virgo Cluster. Cosmic distances are usually judged by 'redshift', the stretching of light from distant objects moving away from us. The Virgo cluster as a whole is moving away from us, but NGC 4216 happens to be moving towards us as the cluster's orientation changes.

40 million light years

2 YOUNG AMONG THE OLD

These Hubble images show fresh star birth in galaxy NGC 4150. Dark strands of dust in the centre provide tentative evidence of a galactic merger long in the past. The smaller inset image shows chaotic activity inside the galaxy's core, with blue areas indicating a flurry of star birth. Stars in the galaxy as a whole are ancient, at about 10 billion years old. The young stars may have formed during an encounter with a smaller galaxy.

44 million light years

3 SHARPEST EDGE

Many spiral galaxies presents themselves edge-on, but few do so with such absolute precision as NGC 5866 in Draco, seen here by Hubble. Its dark spirals of dust make a spectacularly neat line, obscuring the glow of countless stars within the galaxy's core.

44 million light years

4 YOUTHFUL APPEARANCE

Zwicky 18, was thought to be one of the youngest galaxies on record, at just 500 million years old. This youthfulness is typically associated with galaxies that formed in the very early universe, and which, as a consequence, are extremely far away from us. Yet Zwicky 18 is just millions, rather than billions, of light years away. Astronomers are rethinking its age. Perhaps it looks younger than it is.

45 million light years

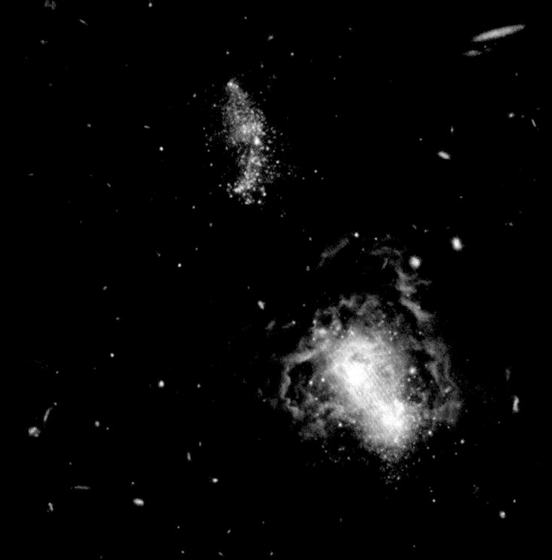

INTO THE DEPTHS OF TIME

When we look into the night sky, we see things as they were long ago, not as they are today. The investigation of deep space is also an exploration of deep time.

Light is the fastest entity in the universe, but it does take time to reach us from the far reaches of space. We see celestial objects as they were when they released the light that has subsequently travelled across the universe and reached our telescopes.

The further away an object is, the further back in time we see it, and unsurprisingly, the more distant an object is, the fainter its light. As we push at the boundaries of time, it becomes that much harder to discern the objects in question, but we can see the dim and tiny glints of light from objects that existed more than thirteen billion years ago, when the universe was a fraction of its current age.

During the past few years, the Hubble Space Telescope has obtained long-exposure images of the faintest objects ever detected. Some of these are galaxies in their early stages, when they were rich in young, hot stars. To look even back through time, we will need a new generation of telescopes that can detect extremely low intensities of infrared light from the faintest and weakest end of the spectrum resolvable to telescopes with mirrors and lenses. Even though that light was emitted in visible energies, it has been stretched by the expansion of the universe, and so appears to us as infrared light.

TALL TOWERS *Edge-on spiral galaxy NGC 3079, located near the constellation Ursa Major, spans some 70,000 light years. Its most significant distinguishing features are pillars of gas, 2,000 light years tall, rising above its centre (50 million light years).*

1, 2, 3, 4, 5 INFLUENTIAL GALAXIES

Even though the galaxies of the Virgo Cluster [1, 2] are tens of millions of light years away, their collective gravitational influence is pulling the Milky Way, and all other galaxies in our Local Group, towards Virgo. The largest in this cluster of some 2,000 galaxies is M87. Bright jets moving at close to the speed of light are seen at all wavelengths coming from the massive black hole at its centre. [3]. Ultraviolet analysis reveals the brightening, over just a few years, of a jet from M87 [4] powered by material heading at near light speed toward one of the most massive black holes yet discovered. Combined data from the Chandra X-ray Observatory and radio telescopes on Earth [5] show that M87's jets lift up relatively cool gas near the centre of the galaxy and produce supersonic shock waves in the galaxy's outer envelope of gas. The process has some physical similarity to volcanic eruptions, where superheated gas and steam rises extremely fast into the cool air above a volcano.

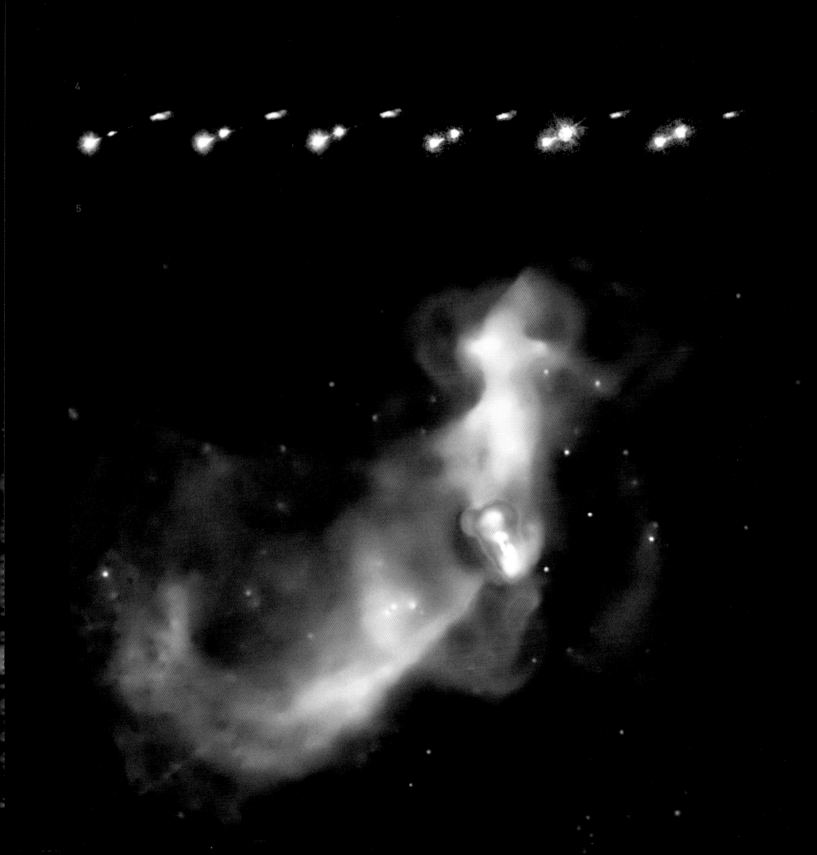

1 Colossal Blast

A monstrous black hole throws huge bubbles of hot gas into space from the hub of the Virgo Cluster galaxy NGC 4438.

50 million light years

2 Supernova Type IA

At first glance, the concentrated fury of galaxy NGC 4526 would seem to be the prime subject of this image. In fact it is the white dot at lower left, the 1994D supernova that is of special interest here. This explosion was not especially different from others of its kind, and that is why it was so useful. By calibrating a precise brightness to distance relationship, astronomers learn more about the expansion rate of the universe.

55 million light years

1

1 REEF OF DUST
Dark filaments of dust have formed a barrier 500 light years deep in this edge-on view of galaxy NGC 4012. The bright sparkles come from newborn clusters of hypergiant stars.
55 million light years

2 GALACTIC CANNIBAL
Astronomical images can be misleading because of the extremely foreshortened perspectives of distant objects. Just for once, this dramatic view is what it seems. The large spiral galaxy NGC 1532 is pulling at the smaller NGC 1531 and will eventually consume it.
55 million light years

2

4, 5 TERRIFYING PHYSICS

This strange spectrum from the Hubble telescope, obtained in 1997, yields compelling evidence for a supermassive black hole at the heart of galaxy M84 [4]. It shows light frequencies emitted from a fast-spinning disc of glowing gas and other material within the black hole's baleful influence, but not yet quite sucked in. The gas is being spun around so fast, the wavelengths of its emitted light appear from our point of view to be stretched into the long red end of the spectrum or compressed into the short blue, depending on whether the material is rushing away from us or hurtling in our direction. What should have been a neat set of spectral lines is now crazily distorted. The image [5] shows the core of the galaxy where the suspected black hole hides.

60 million light years

3 USEFUL YARDSTICK

NGC 4414 is a massive spiral galaxy toward the constellation Coma Berenices. Hubble identified Cepheid variable stars in this galaxy. Cepheids play an important role as a standard measure of luminosity. When we know exactly how bright certain stars should be at near distances, we can calculate the distances of more distant stars.

62 million light years

1 A Scratch in the Dark

Most edge-on views of galaxies show a dark band of dust obscuring the core. This is an exception. NGC 4452 is so thin relative to its diameter that it is actually difficult to determine what type of galaxy it is.

60 million light years

2, 3, 4 Peaceful Merger

The Antennae Galaxies, also known as NGC 4038 and NGC 4039, are a pair of interacting galaxies in the constellation Corvus [2]. They started to interact a few hundred million years ago. Radio emissions [3] betray the presence of a tidal tail of gas and stars, some of which will never be captured into the main populations and will eventually be cut loose to wander the depths of space on their own. A stunning recent image [4] combines data from the Chandra X-ray Observatory in blue, the Hubble Space Telescope in gold and brown, and the Spitzer Space Telescope in red.

63 million light years

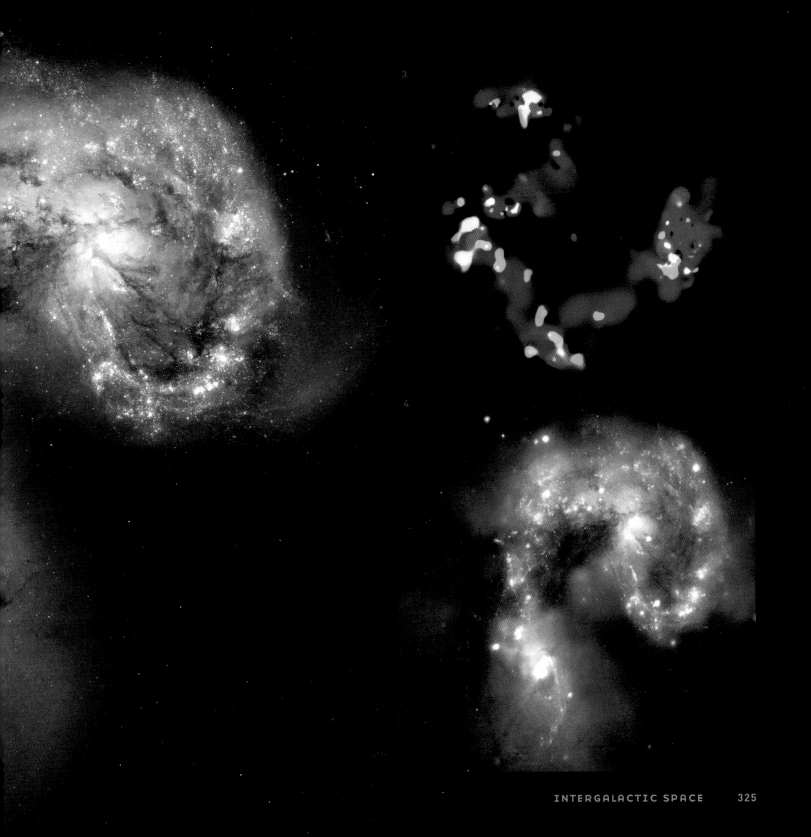

3

4

1 Perfect Bar

This galaxy, NGC 1300 in the constellation Eridanus, is a perfect example of a galaxy with a central bar-shaped structure. The spiral arms begin from the ends of the bar rather than from nearer the centre of the galaxy.

69 million light years

2 Fast Population

Irregular galaxy NGC 1427 is being pulled through space at 600 km per second (370 miles per second) by the combined gravitational force of galaxies in the Fornax cluster. Unable to withstand the strain, the leading edge is breaking away from the tail.

62 million light years

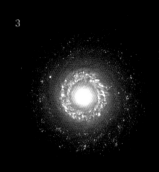

3 Fried Egg

The small spiral galaxy NGC 7742 looks like an egg cooked 'sunny side up'. It is a Seyfert active galaxy, powered by a black hole in its core.

72 million light years

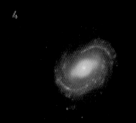

4 Of Little Interest

Galaxy NGC 4579 the constellation Virgo was captured by the Spitzer Telescope's infrared array camera. The red structures are areas where gas and dust are thought to be forming new stars, while the blue light comes from mature stars. In 1779 Charles Messier listed it in his famous catalogue as M58, simply in order not to confuse it with a comet he was tracking.

68 million light years

1 Discrepancy in the Data

A minority of astronomers think that everyone else has miscalculated the size and age of the universe, based on redshift, the stretching of light wavelengths as the relative distances and velocities of objects increase. What if redshift is caused by some other physical factor? Galaxy NGC 4319 appears to be at least 80 million light years way from us, while Markarian 205, also visible here, is more than ten times further away, at one billion light years. The problem is that the two objects seem to be interacting, with a stream of material being drawn from one galaxy to the other, as if they were in the same localized region of space. Something is wrong with someone's numbers.

80 million light years

2 Radio Bright

The lenticular galaxy Fornax A is the fourth-brightest radio source in the sky. Its central region is rotating much faster than the outer shell of stars. As so often with such energetic galaxies, a supermassive black hole is probably driving the dynamics.

75 million light years

1 Diverse Collection

← An ultraviolet image of diverse galaxy types. NGC 3190 is a dusty edge-on spiral, while NGC 3187 is highly distorted. The two are separated by only about half the diameter of our own Milky Way. A ring galaxy, an elliptical galaxy, and other irregular galaxies are also present.

80 million-plus light years

2 Jurassic Spectacle

NGC 1350 is a beautiful spiral galaxy in the southern constellation Fornax . It is approximately 130,000 light years across, slightly larger than the Milky Way. Its spiral arms wind tightly enough to form a prominent central ring. As its light seen here first began the long journey to our telescopes, the dinosaurs were heading towards extinction and the first small mammals were scuttling under their feet.

85 million light years

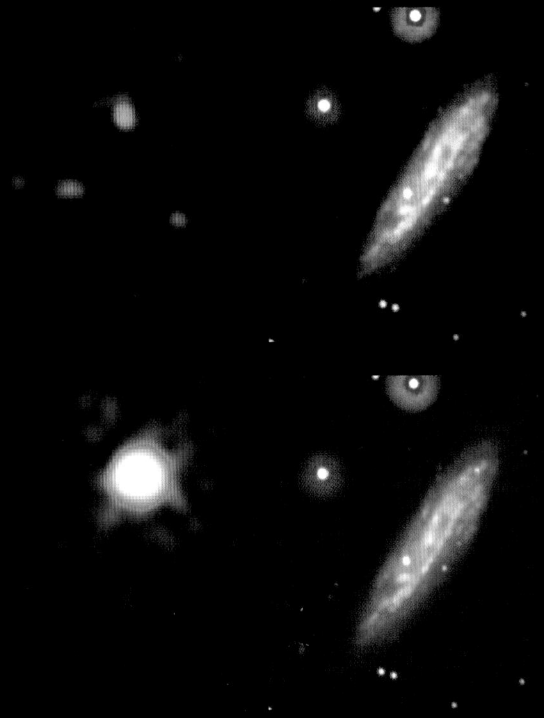

3 Caught in the Act

Over the past century, astronomers have observed thousands of supernova explosions. Typically, a supernova is only noticed some while after the initial explosion. Modern satellites alert astronomers the instant anything explosive is detected. On January 9, 2008, the Swift X-ray Telescope picked up a powerful burst from the distant spiral galaxy NGC 2770. The burst lasted just five minutes before it faded away.

88 million light years

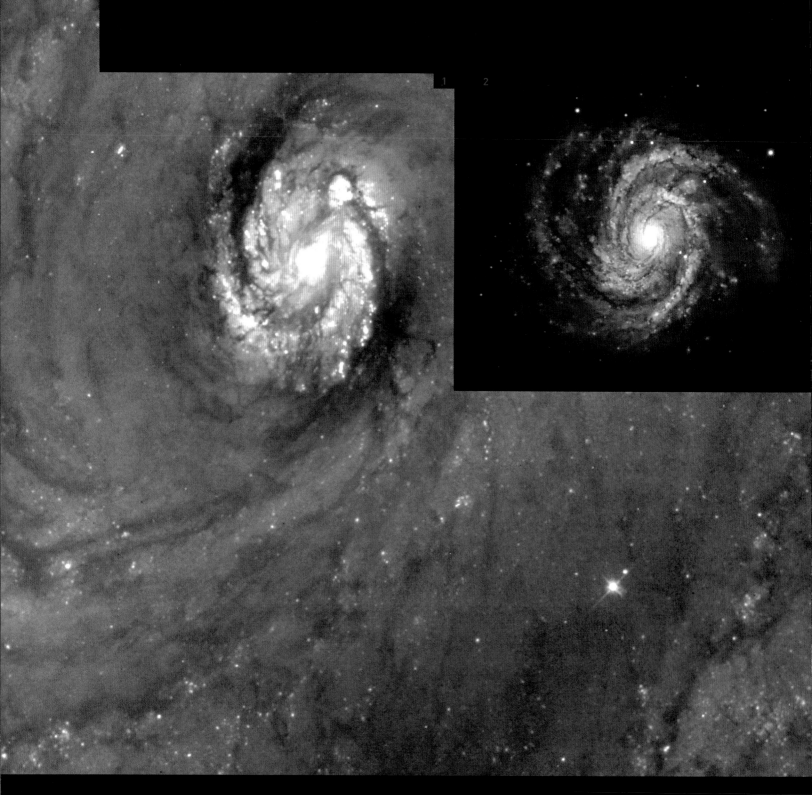

1, 2, 3, 4 DISTANCE MARKERS

Spiral galaxy NGC 3021 [1, 2] hosts a number of Cepheid variable stars pulsating at a rate that is matched closely to their intrinsic brightness. This makes them ideal for measuring intergalactic distances. The Cepheids were used to calibrate an even brighter milepost marker, a Type IA supernova observed in the galaxy during 1995. At 98 million light years distance, the NGC 3370 galaxy [3] nicknamed the Silverado, is also rich in Cepheids. NGC 4603 is 108 million light-years away [4]. Its distance has been accurately measured by astronomers using Cepheids, and thanks to the Hubble Space Telescope's sharp vision, which has picked out 36 Cepheids among this apparent blur of stars.

92 to 108 million light years

1 Incredible Lanes

In this stunning Hubble image of NGC 7049 in the constellation of Indus, a family of globular clusters appears as glittering spots dusted around the galactic halo, but it is the dust lanes that make the image so memorable.

100 million light years

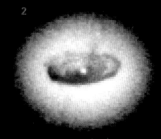

2 Colossal Appetites

The active galactic nucleus of galaxy NGC 4261 contains a black hole with a mass equivalent to 400 million stars. An 800 light-year wide spiral of dust and gas fuels it.

100 million light years

3 Wrong Way Turn

It is easy to think that this galaxy, NGC 4622 in Centaurus, must be rotating counterclockwise. Hubble data surprisingly indicates that the galaxy is probably rotating clockwise, with its outer spiral arms opening outward in the direction of rotation.

111 million light years

1

2

1, 2 Galaxy Battle
The larger galaxy is NGC 2207
and the smaller one, IC 2163.
Strong gravitational forces
have distorted the shape of the
smaller dance partner, flinging
out stars and gas into long
streamers [1] extending 100,000
light years into space. Billions of
years from now, the two galaxies
will become one. Infrared data
from Spitzer [2] show the dark
lanes of dust intermingling.
114 million years

1 NOT A COLLISION

In yet another celestial illusion, galaxies NGC 3314 A and B appear to be colliding with spectacular vigour. It is just an accidental effect caused by our perspective on this distant scene. One galaxy is considerably in the foreground of the other.

117 to 140 million light years

2, 3 RARE GALAXIES

Galaxies with such unusual shapes as these are rare. ESO 350-G40, otherwise known as the Cartwheel Galaxy [2], has a spoked structure, almost certainly the result of a previous collision. The Polar Ring galaxy 4650A [3], located in Centaurus, is probably the result of two galaxies merging, but this explanation for its peculiar structure is not guaranteed to be correct.

130 million light years

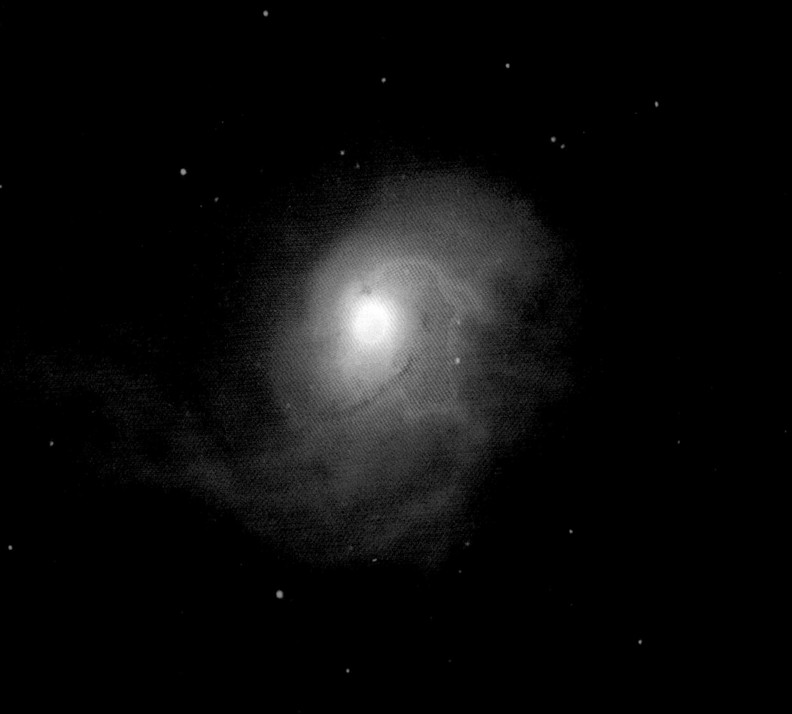

4 Twisted Out of Shape

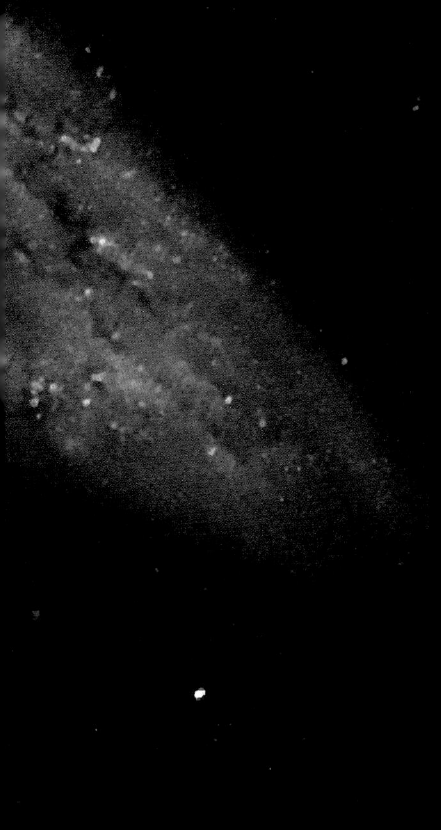

The equatorial dust cloud of ESO 510-G13, a spiral galaxy in the constellation Hydra, is noticeably warped. It has recently collided with another galaxy and is in the process of swallowing it, while rebounding from the gravitational shock.

150 million light years

5 Massive energy

We are accumulating evidence that the energy output of active galaxies, such as NGC 4696, the largest galaxy in the Centaurus cluster, is absolutely colossal, converting mass into energy a hundred times more efficiently than the nuclear fusion inside stars.

150 million light years

1, 2 FUTURE HISTORIES

Several of the dwarf galaxies
in the Hickson Compact Group
31 toward the constellation of
Eridanus are slowly merging [1].
The galaxies will pass through
and destroy each other, millions
of stars will form, millions
of others will explode, and
thousands of nebulae will flare
and dissipate before the dust
settles and a final merged galaxy
wins through in about a billion
years' time. Seyfert's Sextet in
the constellation Serpens is a
similar example of a predictable
merger. The interacting galaxies
[2] are tightly packed into a
region not much larger than the
Milky Way, making the Sextet
one of the densest known galaxy
groups. It, too, will coalesce into
a single large galaxy over the
next few billion years.

170 million light years

3

3 Disrupted Galaxy

NGC 6745A has been disrupted by a near-miss with a galaxy out of frame. A line of new stars trails in the wake of the off-stage interloper.

200 million light years

4 Atoms for Peace

A peculiar elliptical galaxy, NGC 7252 in the constellation Aquarius is thought to be the product of a merger between two disk-shaped galaxies about one billion years ago. Its nickname comes from its prominent loop-like structure, made of stars, that resembles a diagram of an electron orbiting an atomic nucleus.

220 million light years

4

1, 2, 3, 4 MAGNETIC STRUCTURES

Active galaxy NGC 1275, otherwise known as Perseus A, is the central, dominant member of the large Perseus cluster, and a powerful source of x-rays and radio emissions. This stunning visible light image from the Hubble Space Telescope [1] shows galactic debris and filaments of glowing gas, some up to 20,000 light years long. The filaments persist, even though the turmoil of galactic collisions should have destroyed them. Recent studies suggest they are held together by magnetic fields. NASA's Fermi Gamma-ray Space Telescope observed NGC 1275's powerful core as a source of high-energy gamma rays [2] but

the earlier Compton mission did not. The beam from this galaxy's central black hole strengthened in the years between the two missions. Unusual gas filaments [3, 4] are identified from a specific frequency of light emitted by hydrogen, here artificially coloured pink.

235 million light years

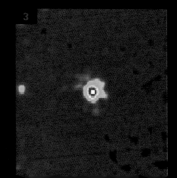

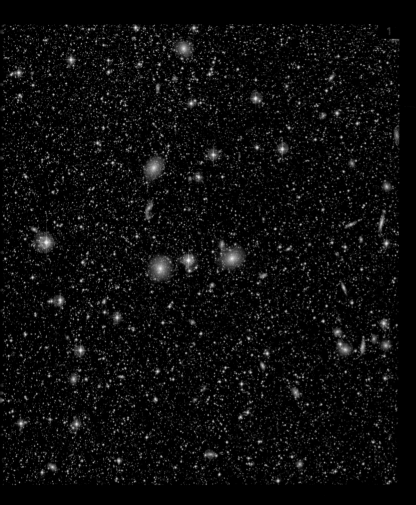

1 GRANDER STRUCTURES

The Norma Cluster, or Abell 3627, makes up about one tenth of the Great Attractor, one of the largest galactic clusters known. As we observe the universe at ever greater distances and larger scales, galaxies become part of far vaster structures formed by the complex interactions of galaxies numbered in the thousands and millions.

250 million light years

2, 3, 4, 5 MULTIPLE PILE-UP

This close view from Hubble [2] reveals galaxies NGC 731A and B at top and 7319 at bottom, with foreground galaxy NGC 7320 partly visible at right. They are part of Stephan's Quintet, a famously tangled pile-up of galaxies in the constellation Pegasus. In a wider view of this region [3], another galaxy, NGC 7331 is prominent at top right. This is one of the few brighter galaxies not included in Charles Messier's famous 18th century listings. Stephan's Quintet is at the bottom left corner. This image [4] shows most of the Quintet at multiple wavelengths, highlighting stellar populations of different ages, while a false colour depiction [5] shows one of the largest shock waves ever seen (the green arc). The wave is produced by one galaxy falling toward another at tremendous speed.

270 million light years

Celestial Tsunami

When galaxies collide at high speed, one can sometimes pass through the other like a bullet, creating a gravitational ripple, as if a stone had plummeted through a galactic pond. This galaxy, AM 0644-741 in the constellation Dorado was shaped by such an impact.

300 million light years

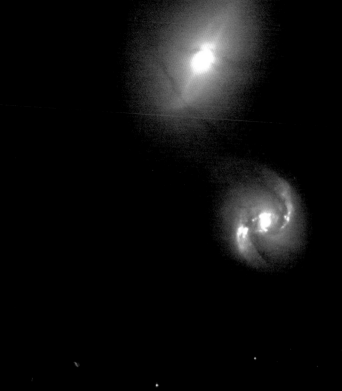

1, 2, 3, 4 PIPELINES, BRIDGES AND TAILS →

Across a span of 20,000 light years, galaxies NGC 1409 and NGC 1410 bridge the gulf between them with tendrils of gas and dust [1], though 'tendrils' may be too delicate a word to conjure up the 40 billion trillion tons of matter that are drawn from one galaxy into the other in the space of a single year. A similarly energetic duel [2] is taking place between NGC 4676 A and B, otherwise known as 'The Mice'. A similar link may not be so obvious in NGC 5257/8, but a pale bridge of stars links them nevertheless. A similar dance [3] is performed by the galactic pair known as Arp 87. Another fine example is the Tadpole Galaxy, also known as UGC 10214 near the constellation Draco [4]. It has been violently disrupted by a collision with a smaller companion galaxy, the compact, blue object in the upper left corner. Strong gravitational forces from the interaction have created a long tail of stars and gas stretching out more than 280,000 light years.

300 million years+

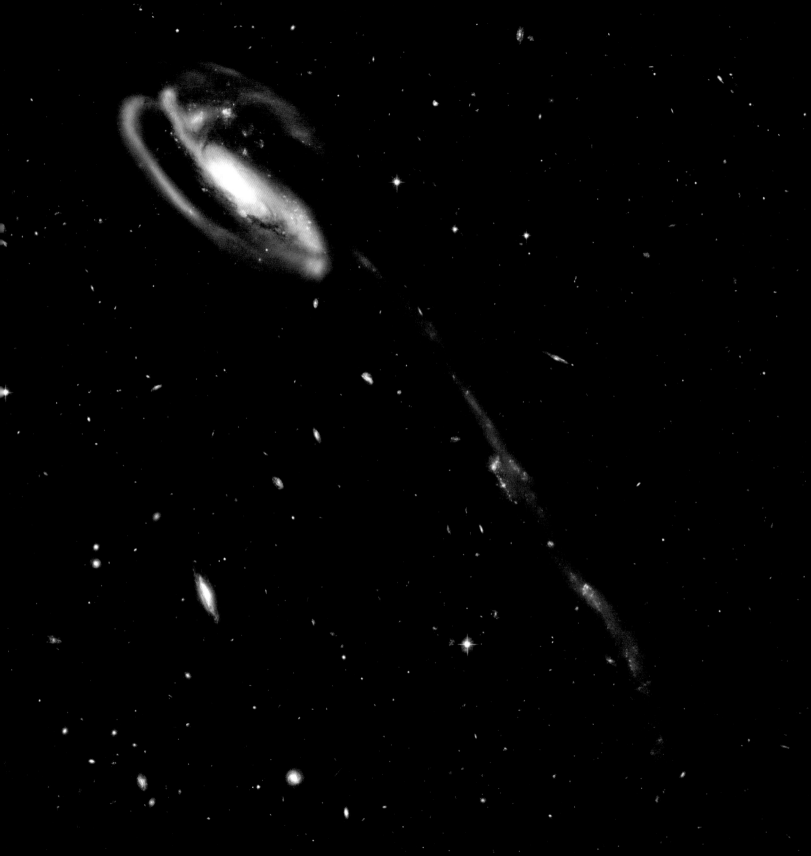

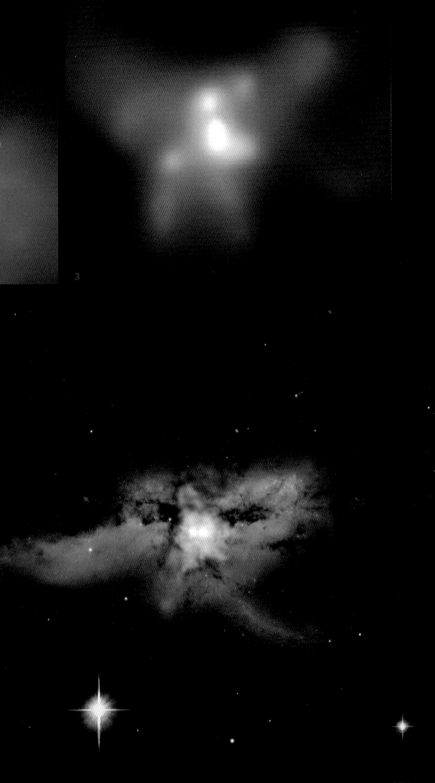

← 1 Fuzzy Swarm

The galaxy NGC 1132 is a cosmic fossil, the aftermath of an enormous galactic pile-up, where the carnage of multiple collisions has resulted in a giant elliptical galaxy far outshining typical galaxies. NGC 1132 is surrounded by thousands of ancient globular clusters, swarming like bees around a hive. These are straggling survivors from destroyed parent galaxies.

300 million light years

2, 3, 4 Supermassive Crashes

Two bright sources at the centre of this composite x-ray and radio image are thought to be supermassive black holes separated by 25,000 light years, each at the core of two merging galaxies in the Abell 400 cluster. Their eventual merger is expected to create intense gravitational waves. This is what such an approaching catastrophe looks like [2] when identified from the monumental x-ray signature. Two black holes in Galaxy 3C75 will clash in just the next three million years' time. These extraordinary multi-wavelength images [3, 4] of NGC 6240 show two black holes [3] a mere 3,000 light years apart. They are the bright point-like sources in the middle of the image.

300 million light years+

1, 2 Gargantuan Galaxy

This seemingly featureless smudge of light [1] is NGC 1132, a galaxy containing a thousand billion stars. An x-ray investigation [2] shows an even vaster shell of gas under the galaxy's influence. Notice the yet more distant galaxies in the background are little more than smudges, even under the Hubble's powerful gaze.

318 million light years

3, 4, 5 Losing Scale →

As we peer yet deeper into the universe, scale values no longer have any human meaning. The Coma Cluster, or Abell 1656, is a cluster of galaxies [3] with over a thousand members. It takes its name from the constellation Coma Berenices. Long-exposure Hubble Telescope images [4, 5] show spiral galaxies NGC 4911 and 4921, within the cluster.

320 million years

1 BEAUTIFUL CONJUNCTION

Four galaxies in the Hickson Compact group HCG 87 are so pleasingly arranged, they look like an illustration of different galaxy types in an astronomy book.

400 million light years

2 IMPERFECT TEN

Hubble aimed its Wide Field camera at this entertaining target, a pair of gravitationally interacting galaxies called Arp 147. They happen to be oriented so that they appear to form the number '10'. The left-most galaxy, or the 'one' in this image, is relatively neat. The right-most galaxy, resembling a zero, exhibits a ragged blue ring of intense star formation.

400 million light years

3 THE HERCULES CLUSTER

Clusters of galaxies are more or less similar in character throughout the universe, but this one is unusually rich in well-formed spiral galaxies. Perhaps this is because the Hercules Cluster, otherwise known as Abell 2151, has experienced relatively few messy collisions. This view is 15 million light years across.

470 million light years

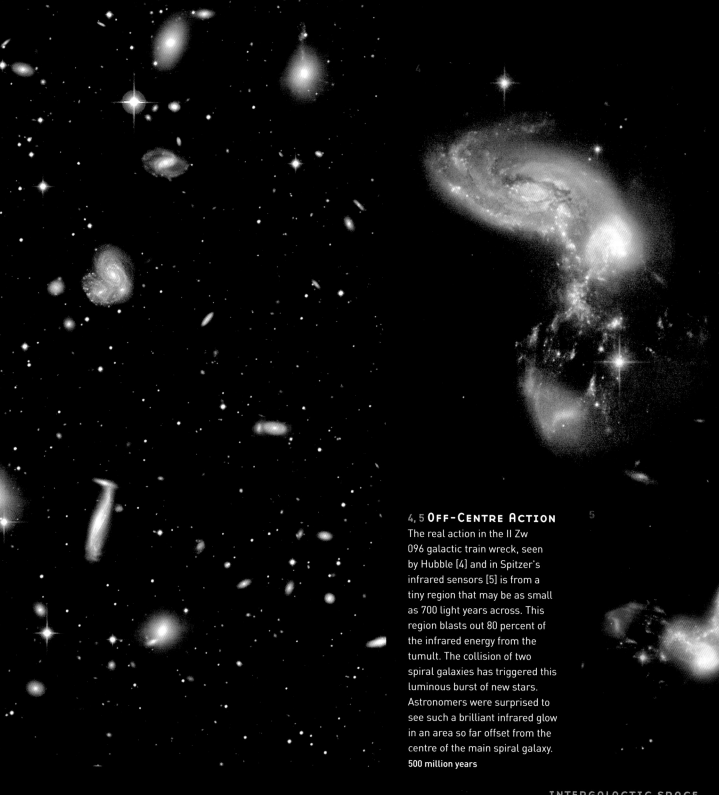

4

4, 5 OFF-CENTRE ACTION

5

The real action in the II Zw 096 galactic train wreck, seen by Hubble [4] and in Spitzer's infrared sensors [5] is from a tiny region that may be as small as 700 light years across. This region blasts out 80 percent of the infrared energy from the tumult. The collision of two spiral galaxies has triggered this luminous burst of new stars. Astronomers were surprised to see such a brilliant infrared glow in an area so far offset from the centre of the main spiral galaxy.
500 million years

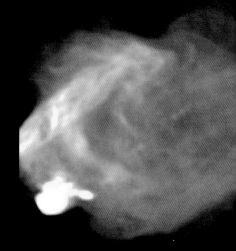

1   **GALLERY OF MERGERS**

Arp 148 is yet another
extraordinary result of an
encounter between two galaxies,
resulting in a ring-shaped galaxy
and a long-tailed companion. The
collision created a gravitational
shockwave that first drew stars,
gas and dust towards the centre
and then caused it to propagate
outwards in a ring.

500 million light years

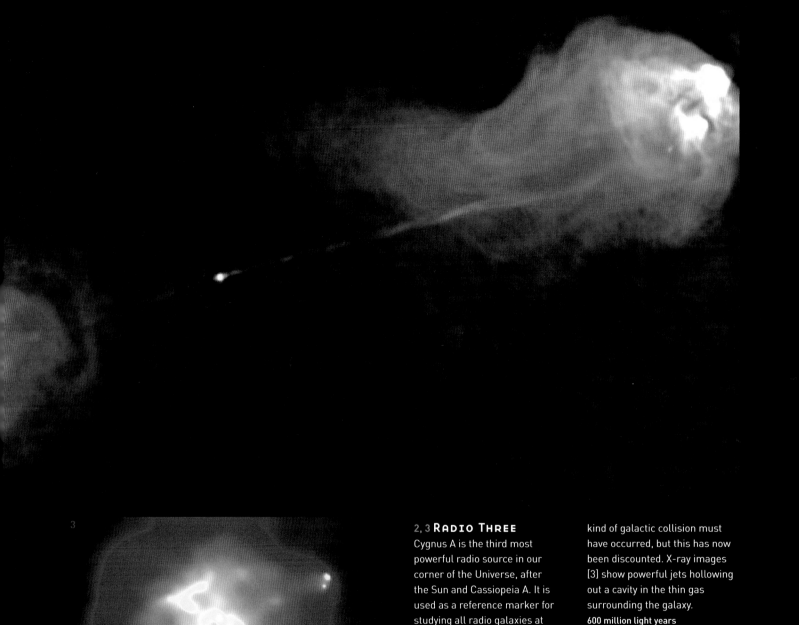

3

2, 3 RADIO THREE

Cygnus A is the third most
powerful radio source in our
corner of the Universe, after
the Sun and Cassiopeia A. It is
used as a reference marker for
studying all radio galaxies at
great distances [2]. The source
of those strong signals is a
mystery. The first photographs of
Cygnus A showed two clumps of
luminous material. Astronomers
naturally thought that some
kind of galactic collision must
have occurred, but this has now
been discounted. X-ray images
[3] show powerful jets hollowing
out a cavity in the thin gas
surrounding the galaxy.
600 million light years

1 PERFECT RING

This is PGC 54559 or Hoag's Object, in the constellation Serpens. It is named after the American astronomer Arthur Hoag who discovered it in 1950. Its unusually neat ring was probably formed in the aftershock of a fast collision with a smaller galaxy, although no obvious trace of that marauder can be found.

600 million light years

2 GREEN GHOST

Hanny's Voorwerp was discovered in 2007 byDutch school teacher Hanny van Arkel, while she was participating as an amateur volunteer in an astronomy project. It appears as a bright blob close to spiral galaxy IC 2497 in the constellation Leo Minor. The galaxy's core produced a quasar, a powerful beacon of energy powered by a black hole. The quasar shot a broad beam in Voorwerp's direction, illuminating the gas cloud. The time lag between the quasar burst and its light reaching the cloud has produced this 'space oddity'. The green glow is a signature of oxygen atoms.

650 million light years

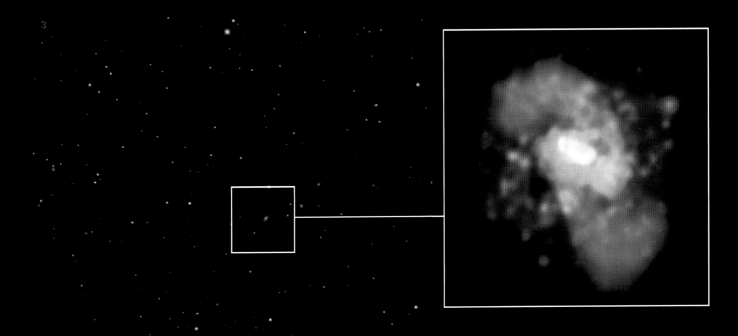

3 DOUBLE FLIP

A giant black hole in radio galaxy 4C +00.58 has been flipped around twice by mergers with other black holes, causing its spin axis to point in a different direction. The smaller image to the right shows a close-up view of this galaxy in x-rays (in gold) from the Chandra X-ray Observatory, and radio waves (in blue) from the Very Large Array radio telescope on Earth.

780 million light years

4 WHO CAN TELL?

This is a rare alignment between two spiral galaxies found in the vicinity of NGC 253, the Sculptor Galaxy, and listed as 2MASX J00482185-2507365. The outer rim of a small, foreground galaxy is silhouetted in front of a larger companion. Skeletal tentacles of dust stretch beyond the small galaxy's disk of starlight. Astronomers have not as yet calculated the distance between the two galaxies, although they think the two are relatively close. At these immense distances, fine-grained determinations to within a few tens of thousands of light years become almost impossible.

800 million light years

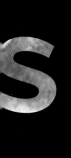

In 1929, the American astronomer Edwin Hubble discovered that all the billions of galaxies in the universe are moving away from each other at great speed.

The Hubble Space Telescope was named in honour of this astronomer, who made some of the most important discoveries in the entire history of astronomy, including the simple and stark fact that numerous distant, faint clouds of light in the universe are entire galaxies. The realization that the Milky Way is only one of many galaxies utterly transformed our understanding of the universe.

The light from distant galaxies stretches into longer wavelengths at the red end of the spectrum, just as sound waves from a speeding train lengthen when the train hurtles past, causing that familiar, mournful lowering of pitch. The greater the distances between galaxies, the faster they fly apart. This 'redshift' is now our best indicator of distances between our galaxy and its farthest relatives. The entire universe is expanding in all directions.

Three years after Hubble's discoveries, a scientifically trained Belgian-born Catholic priest, Georges-Henri Lemaître, alarmed many astronomers with an extraordinary idea, simply by imagining Hubble's expansion time-reversed. Galaxies moving away from each other must once have been closer together. Running the cosmic film backwards to its first frame, so to speak, they must once have been crowded together into the same infinitesimally tiny space. 'We could conceive the beginning of the universe in the form of a unique atom, the atomic weight of which is the total

mass of the universe', Lemaître wrote. He speculated about a 'super-radioactive process' causing that primordial 'atom' to expand into the universe we know today. As other astronomers began to think along the same lines, the concept of the Big Bang was born. But was it just a theory?

In 1965 two radio astronomers, Arno Penzias and Robert Wilson of Bell Laboratories in New Jersey, were testing a sensitive antenna as part of a project to try and identify types of radio noise that can interfere with communications satellites. They wanted to calibrate their antenna against a zero-strength signal by pointing it at

Gravitational lensing around Abell 370 causes light to spread out along multiple paths. The galaxy at right appears crazily distorted. In this cluster, what is actually a single distant quasar 10 billion light years way appears five times.

We have now charted the universe's history to within a few billionths of a second of its birth, when all its matter and energy was concentrated within an infinitely dense and super-hot point, a singularity, smaller than a grain of sand. Within a few more fractions of a second, the furious energies contained in that speck blew it outwards in a gigantic explosion. From this inconceivably powerful cataclysm emerged everything that exists today, and all of space and time, too. After just three seconds, the basic components of matter condensed out of the brilliant storm of energy: the electrons, protons, neutrons and other subatomic building blocks from which the materials of our familiar world are constructed.

In the wake of these swift and dramatic events, the next 10,000 years were relatively uneventful. The universe was still too hot for anything much to happen except for the continuing, remorseless expansion into a larger and larger space, which that same expansion actually created as it progressed. Fierce radiation dominated this phase, preventing any of the matter particles from condensing into atoms.

After 300,000 years, as the fireball at last began to cool down, the subatomic particles were able to link together into complete atoms of hydrogen, helium and lithium, the three lightest and simplest chemical elements. At last, after three hundred million years, the first distinct physical structures emerged as vast clouds of hydrogen and helium coalesced under the force of gravity. These regional variations were the seeds of future galaxies. At smaller scales, tightly localized swirls of gas were drawn together into hot, dense spheres. The gravitational forces and internal pressures increased even further, and the spheres reached a critical point, igniting into the first generation of stars.

narrow windows of sky where the density of stars or distant galaxies is so low that no radio noise could possibly be detectable. They were vexed that a constant low level hiss of microwave background noise refused to go away. They checked and rechecked all their equipment, and even climbed inside the dish of their radio antenna to evict a pair of pigeons. They tried everything, but the irritating background interference continued. They moved the antenna around the sky, looking for other quiet spots, only to discover that there are no quiet spots. That annoying hiss comes from everywhere in the sky. It is the faint echo of Big Bang.

THE BIGGER PICTURE

All that we see of the universe, even through our most sensitive instruments, turns out to be just a fraction of what is really out there.

The total gravity of the universe is far greater than the mass of all its galaxies can explain, while individual galaxies contain much more mass, or 'dark matter', than all the stars, planets and gas clouds within them. Dark matter may consist of subatomic particles that possess mass, and therefore create gravitational effects, but which otherwise hardly influence the matter that we can observe. Dark matter and its related phenomenon, dark energy, account for nine tenths of the universe. What we see in our instruments is just the remaining one tenth that we know how to investigate.

Light from distant galaxies is bent by the gravity of foreground galaxies. If current theories are right, most of that gravity is generated by dark matter. Astronomers have created maps of dark matter distribution. They are only simulations, not pictures. We cannot directly see dark matter.

The rate of the universe's expansion is also increasing under the influence of dark energy. Then there is dark flow. At least 700 galactic clusters are being pulled toward a region between the constellations of Centaurus and Vela. This motion is different from the general expansion of the universe, and cannot be explained by any known forces. Whatever is responsible may lie outside the observable universe.

AN ILL WIND *The larger of two galaxies in 3C 321, dubbed the 'death star galaxy' by astronomers, shoots an immensely powerful jet from around the black hole at its centre. The unfortunate smaller galaxy has apparently swung into the jet's line of fire, and is being sprayed with energetic particles and magnetic fields. Jets produce staggering amounts of radiation, especially in the form of dangerous x-rays and gamma-rays. At only about 20,000 light-years apart, the two galaxies of 3C 321 are about the same distance apart as the Earth is from the centre of the Milky Way. Any planets caught in the jet spanning that gulf will be swamped in radiation. (1.4 billion light years).*

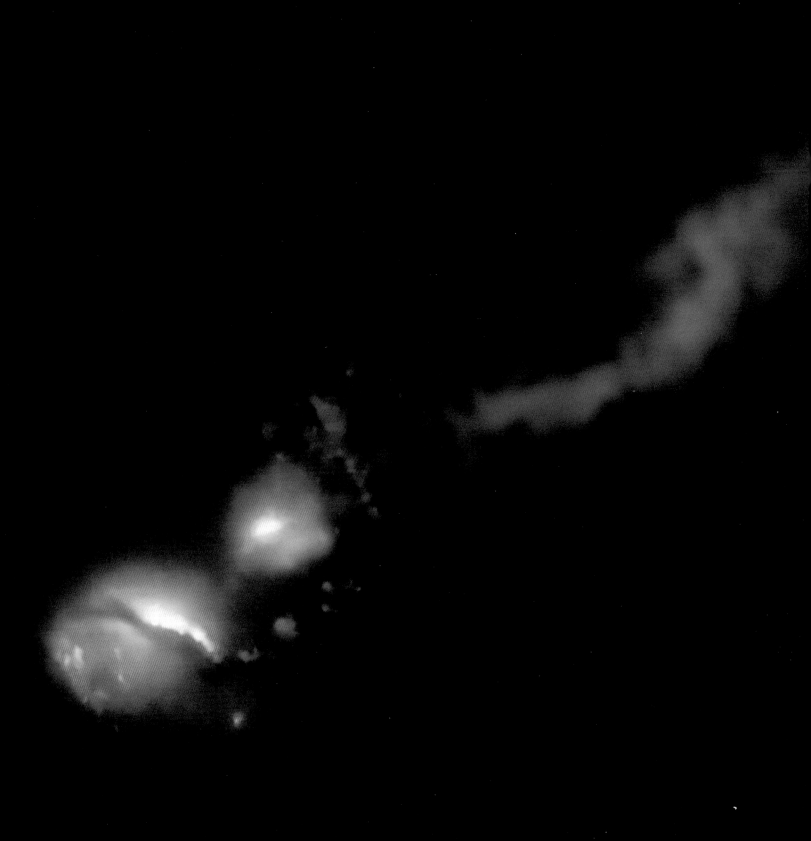

Just Scratching the Surface

This all-sky survey maps the positions of 1.6 million galaxies within 1.3 billion light years of the Milky Way. Just as stars are sprinkled like dust in galaxies, so galaxies are sprinkled like dust across the sky. Yet this is just a tiny fraction of the total, because this survey reaches outwards to only one tenth of the observable limits of the universe.

1.3 billion light years

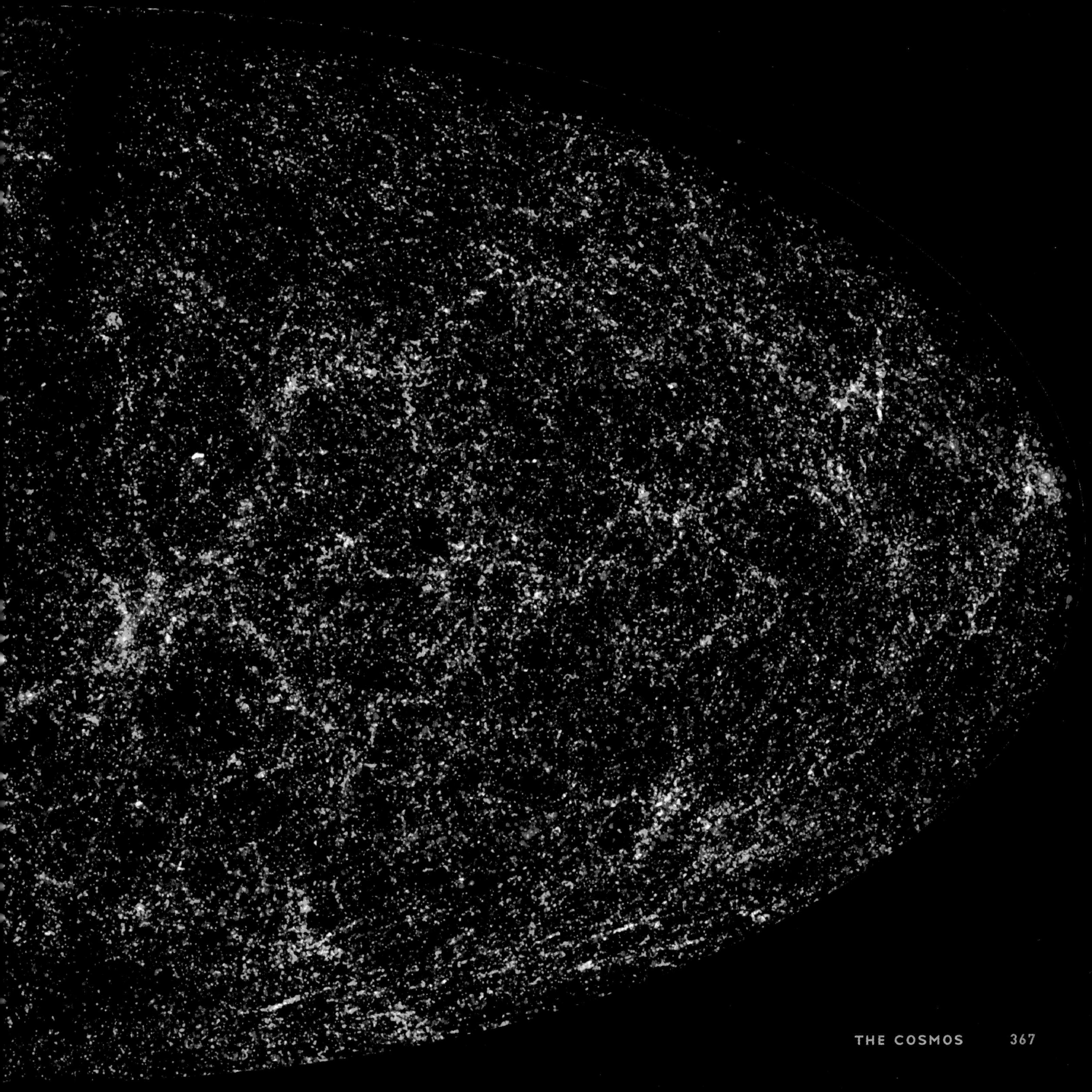

1, 2 HOMELESS QUASAR
NASA's Hubble Space Telescope and the European Southern Observatory's Very Large Telescope (VLT) at Cerro Paranal in Chile combined forces to study this bright quasar HE1239-2426. Quasars are powerful and typically very distant sources of radiation commonly associated with galaxies that have a central black hole, yet this one [1] seems to lack a surrounding galaxy. It may be a rare case of collision between a normal galaxy and a more exotic object, as yet unknown, harbouring a massive black hole. Quasar HE1239-246 was thought to be similarly lacking a galaxy, until Hubble finally resolved the faint light from its surrounding galaxy [2].
1.5 billion light years

3, 4, 5 SHAPING SPACE

Galaxies create the shape of the space that they inhabit. This cluster, Abell 1689 contains hundreds of galaxies and many trillions of stars. It visibly warps space, 'lensing' background galaxies [3] as their light bends around the sides of the foreground group. But the galaxies alone do not contain sufficient mass to explain the curvature of space seen here. 'Dark matter' contributes nine tenths of the universe's entire mass. Hubble's observations of Abell 1689 were used to produce this map of dark matter distribution in and around the cluster [4]. A similar gravitational lensing effect is seen in Abell 2218, where multiple images are created of a very young galaxy some 13 billion years further in the distance [5]. It can be seen as the faint smears at the left of this image.

2.2 billion light years +

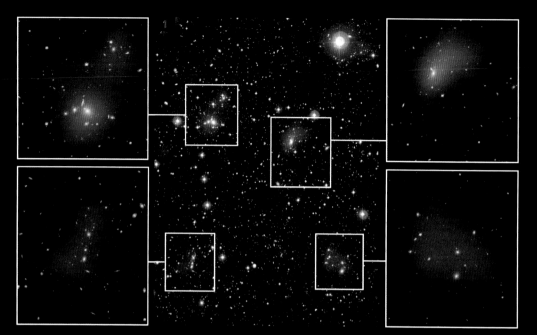

1 DARK MATTER

Supercluster Abell 901/902 is composed of hundreds of galaxies. Astronomers assembled this composite of visible light images taken by ground-based telescopes, and a dark matter map derived from Hubble observations. The magenta-tinted clumps represent the dark matter, an invisible form of matter that accounts for most of the universe's mass. Hubble cannot see the dark matter directly. It is calculated from the effects of gravitational lensing.

2.6 billion light years +

2 MORE DARK EVIDENCE

Galaxy cluster 1E 0657-56, is also known as the 'bullet cluster'. Individual galaxies are seen in the optical image data, but their total mass adds up to far less than the mass of the cluster's two clouds of hot x-ray emitting gas, shown in red. Representing even more mass than the optical galaxies and x-ray gas combined, the blue regions show the distribution of dark matter in the cluster.

3.5 billion light years

3 CLUSTER WITH BLACK HOLES →

The Chandra X-Ray Observatory has examined the most powerful eruption ever seen in the universe, coming from the heart of distant galaxy cluster MS 0735.6+7421. The energy is being released by a supermassive black hole that has already sucked in the mass equivalent of a billion stars. Its output is causing the greatest energy burst since the Big Bang.

2.6 billion light years

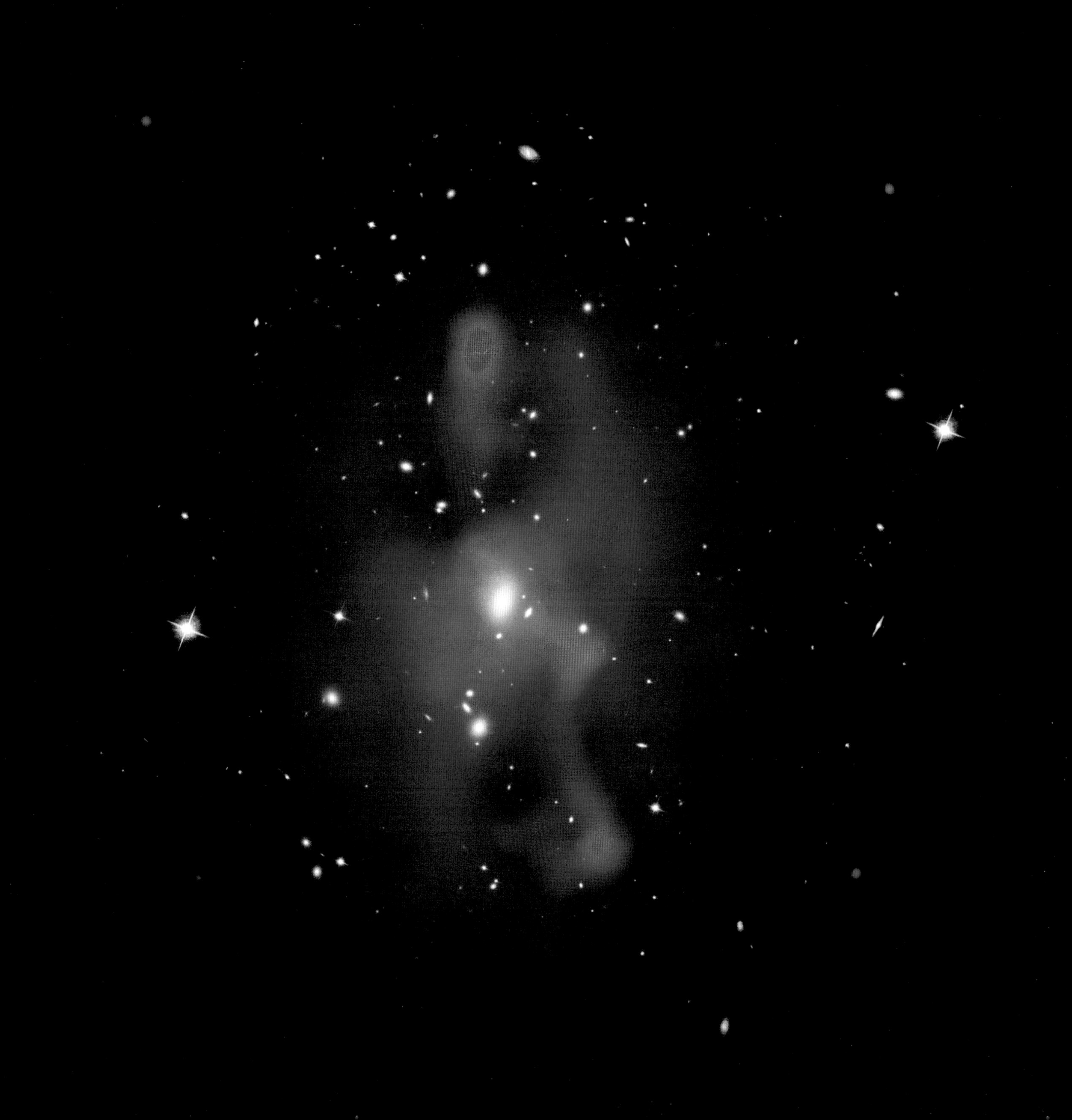

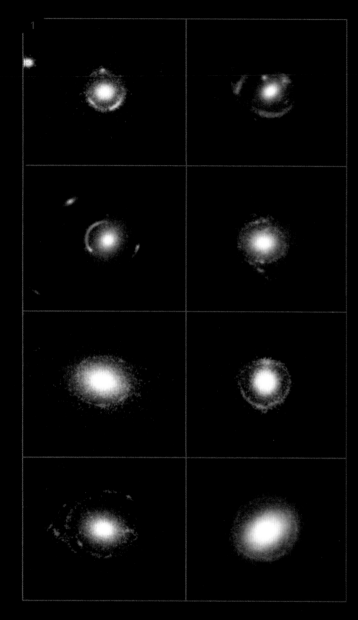

2 First Found

This recent composite from Hubble, Chandra and Spitzer shows a giant jet of particles shooting out from quasar 3C273. The jet is enormous, stretching across more than 100,000 light years, a distance comparable to the span of the entire Milky Way. Quasars are supermassive black holes surrounded by turbulent material, which is heated up as it is dragged toward the black hole. The material glows brilliantly, and some of it is blown off into space in the form of jets. This quasar, identified in 1963, was the first one ever discovered.

3 billion light years

4 Dark Ring →

Observations of gravitational lensing around galactic cluster C10024+17 and subsequent mapping suggests that, contrary to what astronomers expected, dark matter surrounds the cluster in a ring rather than being concentrated within the cluster itself.

5 billion light years

1 Einstein Rings

A selection of giant elliptical galaxies demonstrates the characteristic bull's-eye patterns created when light from galaxies twice as far away is distorted by the gravity of the foreground galaxies. This phenomenon is called gravitational lensing, and was first predicted by Albert Einstein almost a century ago. The bull's-eye patterns are called 'Einstein rings'.

2 billion - 4 billion light years

3 Naked Quasar

HE0450-2958 is one of the most assured instances of a 'naked' quasar. There is a total absence of starlight that would be expected from a host galaxy. The cloud of gas, visible at top left, is glowing in the intense radiation coming from the quasar. Most likely, it is the gas that feeds the quasar's black hole. The bright object at lower right is an incidental foreground star.

5 billion light years

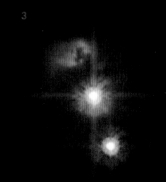

5, 6 PERFECT DISTORTION

Perfect Distortion In Einstein's Ring SDSSJ0946+106, the gravitational field of an elliptical galaxy warps the light of two galaxies exactly behind it [5]. The massive foreground galaxy is almost perfectly aligned in the sky with two background galaxies at six billion and eleven billion light years distance respectively. A close-up of the Einstein ring reveals further detail [6], including a faint secondary ring. Such precise alignments are extraordinarily rare.

2, 3 UNRELIABLE EXTREMES

As we head towards the limits of the optically observable universe, the 'pictures' that we see [3] are increasingly less reliable guides to the shape of celestial objects. Gravitational lensing around Abell 370 causes light to spread out along multiple paths. The galaxy [2] appears crazily distorted. In this cluster, what is actually a single distant quasar 10 billion light years way appears five times.

7+ billion light years

1 COLOSSAL CRASH

One of the biggest galaxy collisions ever observed is taking place at the centre of this image. The four yellow blobs in the middle are large galaxies that have begun to tangle, and which will ultimately merge into a single gargantuan galaxy. The yellowish cloud around the colliding galaxies contains billions of stars hurled out during the messy encounter.

5 billion light years

4 Gamma-ray burst

Of all the mysteries in the universe, gamma-ray bursts are among the greatest, in terms of power and peculiarity. For a brief moment in January 1999, a burst logged as GRB 990123 shone with the equivalent brightness of 100,000 trillion Suns. Fortunately, this epic blast was two-thirds the way to the edge of the known universe. Gamma-ray bursts may be the result of black hole collisions, but the truth is that the cause of these exceptionally distant and unimaginably powerful blasts is still unknown.

9 billion light years

THE MYSTERIOUS UNIVERSE

Is this universe the only one? Was the Big Bang a unique event, or just one of many? Could black holes be linked to other universes?

The space-twisting forces and inconceivable gravitational energies inside black holes are so extreme, they reach beyond our current understanding. The 'other side' of a black hole may be a 'white hole' giving rise to another universe with its own space-time dimensions, completely separate from ours. The Big Bang from which our universe was born was a singularity, a dimensionless point of infinitely compressed matter. Black hole singularities also may be dimensionless points of infinitely compressed matter. The two singularities, Black and Bang, could perhaps be equivalent.

Are black holes responsible for creating other universes? Is such a thing even possible? Many exciting ideas are taking shape at the cutting edge of cosmology and theoretical physics. We may be completely wrong, but perhaps this universe is merely part of an even grander 'multiverse'.

We also do not know if the universe that we see will persist forever, or if it will come to an end at some time in the far future. We suspect that all its energy finally dissipates too thinly for anything interesting to happen, such as star and planet formation. To put it simply, perhaps the universe eventually burns itself out, leaving only cold, thin dust. Again, we do not know. And that is why astronomy and cosmology will continue to excite us for generations to come.

DARK MATTER *The as-yet unidentified material that accounts for nine tenths of the universe's total mass. Along these tangled skeins and webs of gravitational influence, normal matter accumulated as vast collections of galaxies, galactic clusters and superclusters.*

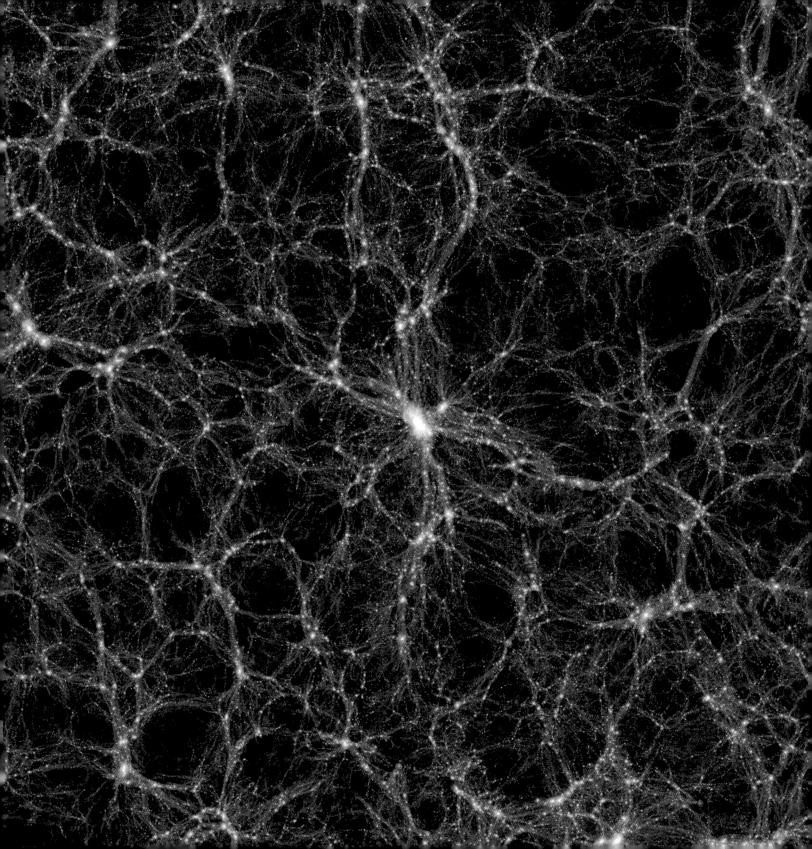

3 Light from 1.6 million galaxies mapped

We are only beginning to perceive the larger-scale structures of the universe. Many galaxies are gravitationally bound together to form clusters, which themselves are loosely bound into superclusters, which in turn are sometimes seen to align in even larger structures.

1 Mature For Their Age

Redshift requires distant galaxies to be detected in longer wavelengths than nearer galaxies. We might expect the populations to thin out as we see ever further back into the universe's history, but a surprisingly large collection is still evident in this composite image combining infrared and visible light surveys. Astronomers were surprised to find such 'modern' galactic clusters in an era when galaxies were thought not to have had sufficient time to reach such sizes.

10 billion light years

2 Farthest Cluster

This fuzzy blob is JKCS041, the most distant and therefore the youngest galactic cluster yet observed. We see the cluster today as it appeared when the universe was less than a quarter of its present age.

11 billion light years

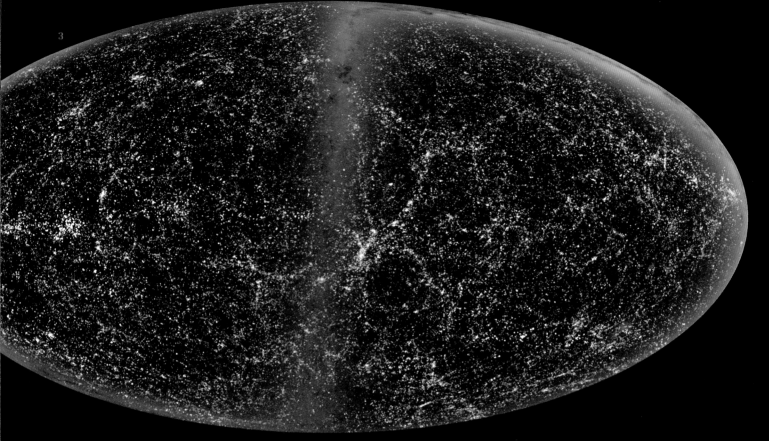

4, 5 Galaxies Like Grains of Sand

Thousands of galaxies crowd into this Herschel image of the distant universe [4]. Each dot is an entire galaxy containing billions, or in many cases, trillions, of stars. The mottled effect in the image gives away the tendency of galaxies to cluster. This patch of sky covers around 60 times the apparent size of the Full Moon. It was taken in a region of space called the Lockman Hole, near Ursa Major, which allows a clear line of sight through gaps between nearby stars and out into the distant universe. This scan [5] of a tiny fragment of sky unites extremely deep observations from Spitzer and Hubble. Blue and white galaxies are sites of rapid star formation in the relatively nearby Universe (though the nearest galaxies in the field are more than a billion light years away). Orange galaxies are far more distant, while faint red blobs are the most distant of all.

10–12 billion light years

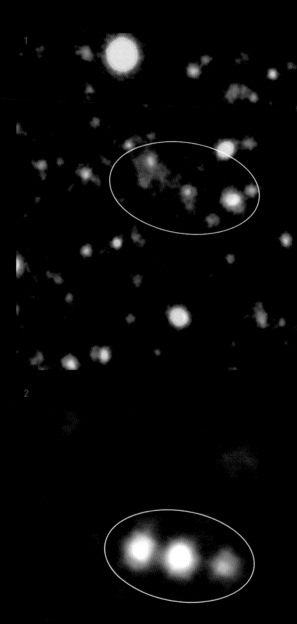

1, 2 BLOB GALAXIES

This composite from the Spitzer telescope shows a giant cloud of hot hydrogen gas [1], and three monstrously bright galaxies [2], trillions of times brighter than the Sun, in the process of merging together, seen here in yellow. The envelopes of gas are about ten times as large as the galaxies they surround.

11 billion light years

3

3, 4 VERY YOUNG GALAXIES

The Spitzer and Hubble telescopes combine to show HUDF-JD2, one of the youngest and most distant large galaxies ever seen. It dates from just 800 million years after the Big Bang. The Milky Way by comparison is approximately 13 billion years old. A blow-up of a small area of the Hubble Ultra Deep Field image [4] at top right shows where the galaxy is located, inside the green circle. At centre right, the galaxy is seen in Hubble's near-infrared camera. The final image at bottom right, from Spitzer's infrared camera, shows the light from older, redder stars in this incredibly distant object. This Hubble collage [3] shows nine very compact, dense and young galaxies as they appeared 11 billion years ago. They are typically only about 5,000 light-years across, yet each contains 200 billion times more mass than the Sun. They are a fraction of the size of today's grown-up galaxies but host the same number of stars.

13.9 billion light years

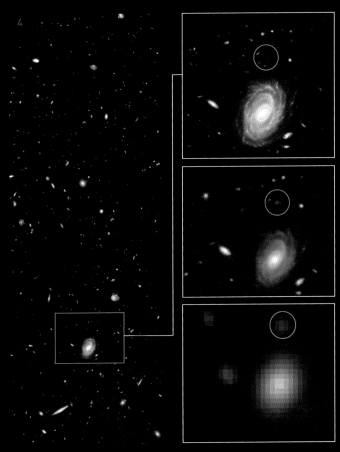

4

1 Juvenile Tadpoles

Another collection shows young galaxies caught in the act of merging. All of them appear as they existed many billions of years ago. Astronomers have dubbed them 'tadpole galaxies' because of their distinct knot-and-tail shapes. They are considerably smaller than today's giant galaxies.

13 billion light years

2 X Marks the Burst

Chandra monitors of the afterglow of gamma-ray burst GRB 020813, revealing an abundance of chemical elements ejected by the supernova explosion of a massive star. Narrow lines and bumps in the spectrum betray the presence of different atomic elements moving away from the site of the burst at a tenth the speed of light, probably as part of a shell of matter ejected in the supernova explosion. The data supports the theory that particularly massive supernovae may create gamma-ray bursts. The collapse of a massive star's core precipitates an explosion that ejects the outer layers at high speeds, while the collapsed core forms a black hole surrounded by a disk of debris. A short time later, this black hole-disk system produces a jet of high-energy particles.

9 billion years

3 At the Limits of Redshift

These are among the earliest and therefore youngest dwarf galaxies that Hubble can detect. In reality their young stars shine hot and blue, but redshift has dragged their light towards the cooler end of the spectrum.

13 billion light years

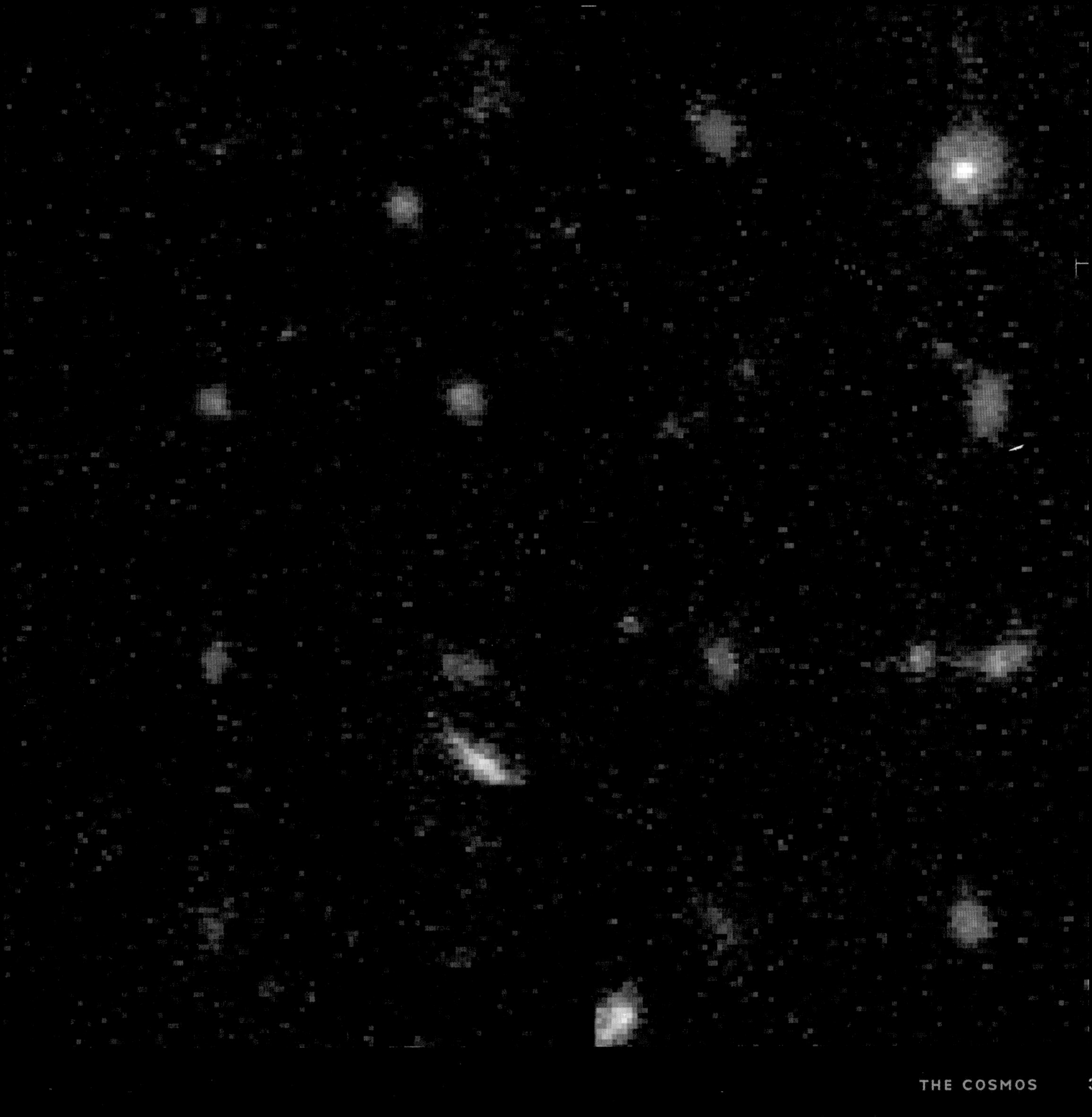

HUBBLE ULTRA DEEP FIELD

The Hubble Ultra Deep Field cuts across 13 billion light years to capture some of the most distant galaxies known, formed when the universe was under 800 million years old. These, the smallest and reddest, are pictured alongside a myriad of 10,000 younger galaxies. Their often unusual and irregular forms chronicle an era when order was just beginning to emerge from chaos.

13 billion light years

1, 2 Earliest Explosions

This Gamma-ray explosion happened just 630 million years after the Big Bang. GRB 090423 detonated so early that astronomers have no evidence that anything capable of exploding could have existed any earlier. The afterglow [1] is circled in the image, taken by the large Gemini North Telescope in Hawaii. One possibility is that the burst occurred in one of the very first generation of stars, announcing the birth of an early black hole. The so-called Cloverleaf Quasar [2] looks like the product of multiple black holes, but is in fact a gravitationally lensed image of just one quasar, again dating from a very early era in the universe's history.

13 billion light years

3, 4 Superclusters

Further analysis of the microwave sky by the orbiting Planck and XMM-Newton observatories produced the very first images [3, 4] of superclusters, amongst the largest objects seen so far in the universe. They form at the intersections of dark matter filaments. The 'Sunyaev-Zel'dovich effect' is a characteristic signature that they imprint on the Cosmic Microwave Background. This supercluster comprises many thousands of galaxies.

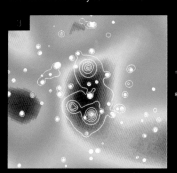

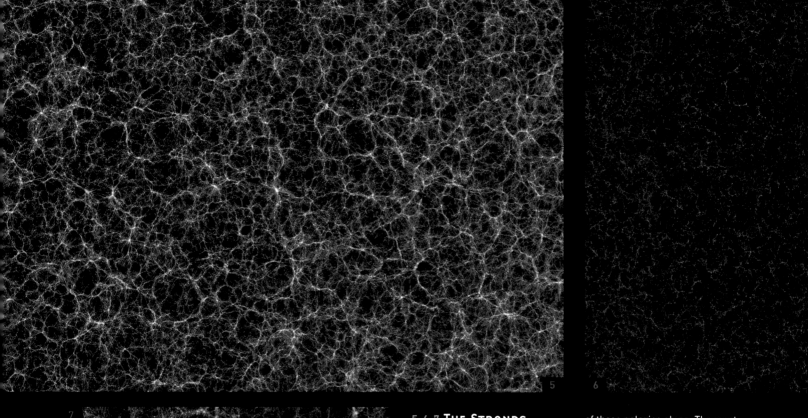

5

6

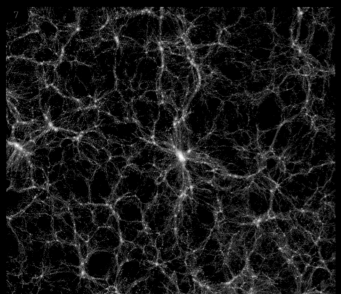

7

5, 6, 7 THE STRANDS OF CREATION

Dark Matter, the as-yet unidentified material that accounts for nine tenths of the universe's total mass, is calculated from the gravitational influences that galaxies and galactic clusters exert, yet which cannot be explained by the mass of those galaxies alone. The most powerful computers are required to extrapolate data and map the likely pattern of dark matter. Along these tangled skeins and webs of gravitational influence, normal matter accumulated [6] as vast collections of galaxies, galactic clusters and superclusters [5, 7].

MAPPING BIG BANG

One of the great achievements in modern science has been the discovery and mapping of the microwave echo of Big Bang, the last, faint traces of the explosion of matter and energy that gave rise to the universe. This map [1] is derived from the Wilkinson Microwave Anisotropic Probe, or WMAP. This satellite scanned the entire sky looking for the distribution patterns of the echo, known as the Cosmic Microwave Background, or CMB. Random variations in the density of the universe during its first few fractions of a second have expanded ceaselessly over 13.75 billion years. If those fluctuations, no larger than quantum shimmers among atoms, had not existed, then the universe would have ballooned perfectly evenly in all directions, and galaxies and stars might never have formed. The WMAP image is processed so that the 'noise' of the foreground Milky Way does not interfere with the sensitive data.

In another iteration using different instruments, the Milky Way Galaxy sprawls across the middle of this false-colour, all sky view [2]. The bright stripe microwave energies in the galaxy are hundreds or thousands of light years away, while the mottled regions at the top and bottom represent the CMB radiation, some 13.7 billion light years distant.

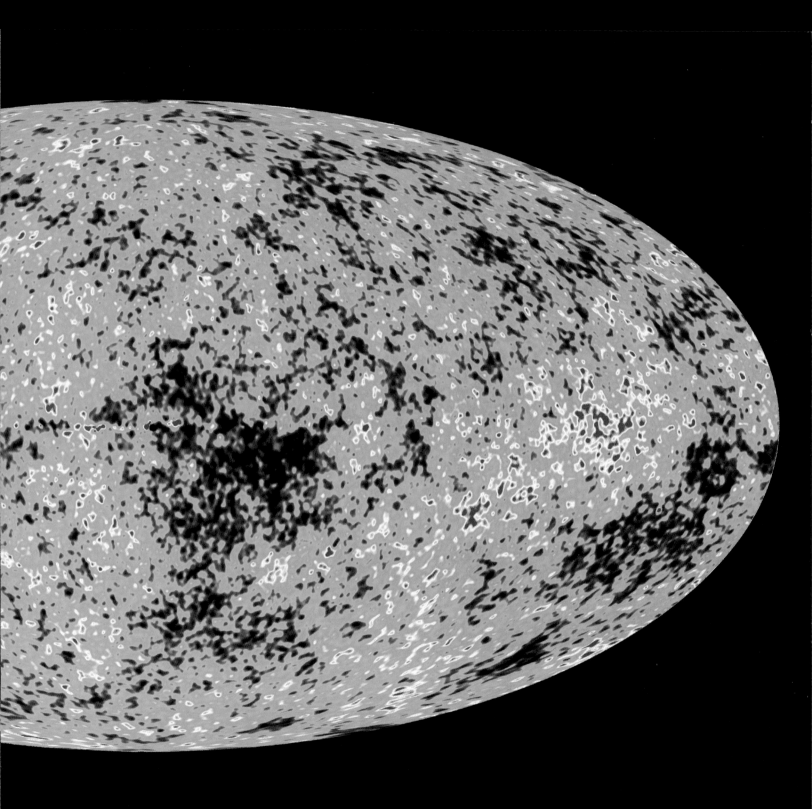

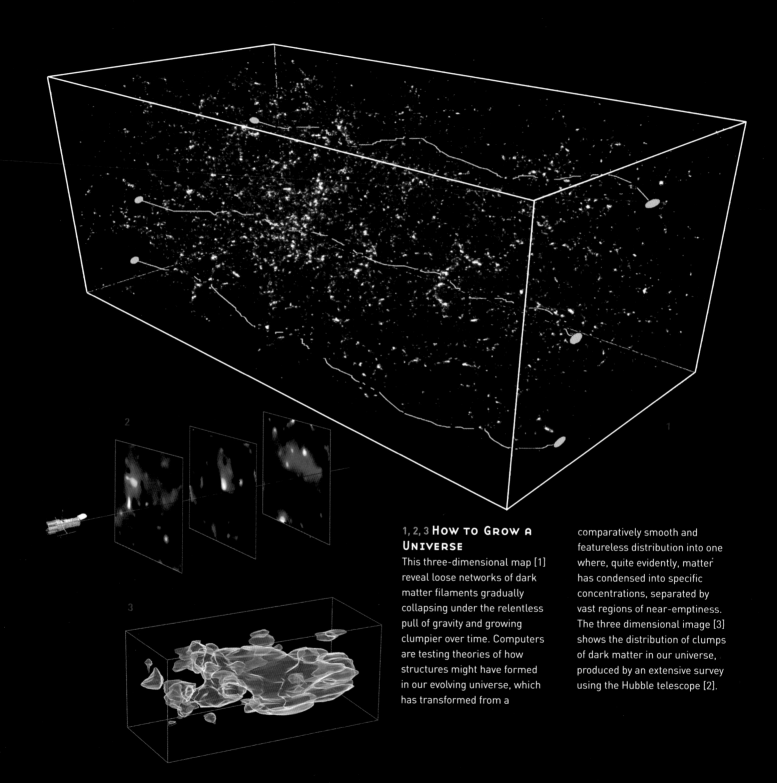

1, 2, 3 How to Grow a Universe

This three-dimensional map [1] reveal loose networks of dark matter filaments gradually collapsing under the relentless pull of gravity and growing clumpier over time. Computers are testing theories of how structures might have formed in our evolving universe, which has transformed from a comparatively smooth and featureless distribution into one where, quite evidently, matter has condensed into specific concentrations, separated by vast regions of near-emptiness. The three dimensional image [3] shows the distribution of clumps of dark matter in our universe, produced by an extensive survey using the Hubble telescope [2].

4 Fast Holes

In the last few years, hundreds of extremely distant galaxies' have been found to contain black holes. Evidently they began to form sooner in the universe's history than astronomers once thought. Creation and destruction have walked hand in hand since the dawn of existence.

11 billion light years

5 Distant Fog

The Herschel space telescope has shown that previously unseen distant galaxies are responsible for a cosmic 'fog' of infrared radiation. The galaxies are from a time when the universe was just 16 percent of its current age.

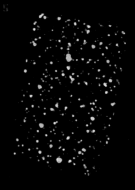

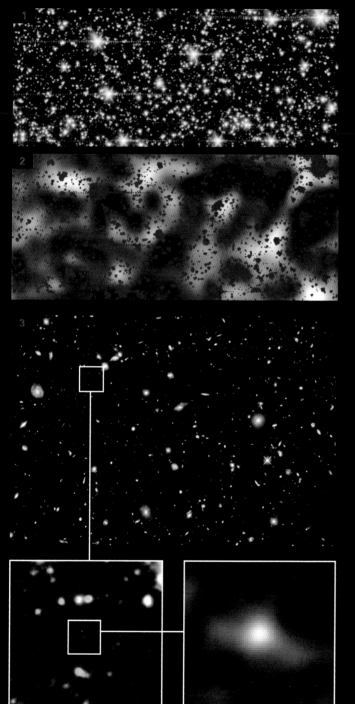

1, 2 Earliest Light

The Spitzer Space Telescope examined visible stars and galaxies in the constellation Draco, covering a region 100 million light years across [1]. After the light from all the obvious stars and galaxies was masked out by digital processing [2] what remains is a faint infrared glow from the first stars in the universe.

3 Most distant

This object's light traveled 13.2 billion years to reach Hubble. The tiny, dim object is a compact galaxy of blue stars that existed 480 million years after the Big Bang. More than 100 such mini-galaxies would be needed to make up our Milky Way. The new research offers surprising evidence that the rate of star birth in the early universe grew dramatically, increasing by about a factor of 10 from 480 million years to 650 million years after the Big Bang.

13.2 billion light years

4 Bent Space

This dramatic image, obtained in December 2010, was one of the first to be produced by e-MERLIN, a powerful new array of radio telescopes linked across the UK, and operated by the University of Manchester's famous Jodrell Bank Observatory. It shows the light from a quasar bent around a foreground galaxy by the curvature of space.

9 billion light years

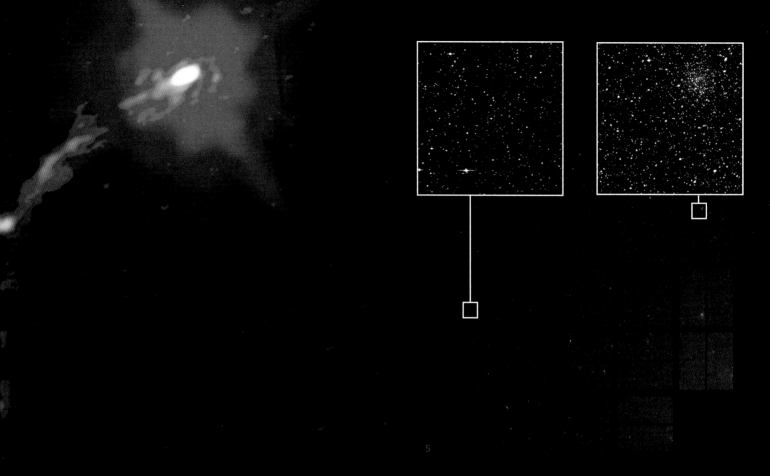

5

5 Looking for the Others

The Kepler Space Telescope kept watch, relentlessly, over 156,000 stars in the constellations Lyra and Cygnus, looking for tiny changes in their light output caused by the shadows of planets transiting across their discs. By February 2011, the tally of potential planets came to 1,235, with at least 170 stars hosting entire solar systems. The grid pattern of vertical and horizontal rectangles [5] represents 42 charge-coupled devices (CCDs) in Kepler's sensor array. A star with a known planet, called TrES-2, is highlighted at the bottom left of one array [detail left]. Its hot Jupiter-like planet transits the star every 2.5 days. The NGC 6791 star cluster detected by another array [detail right] is eight billion years old. These stars were simply test targets prior to Kepler's more detailed search for planets, using its entire resolution capacity of 95 megapixels.

INDEX

PICTURE CREDITS

Key: l = left, **r** = right, **t** = top, **b** = bottom. **m** = middle

11 NASA; **13** NSO/AURA/NSF; **14** Mats Löfdahl, ISP, SST, Swedish Royal Academy of Sciences; **15br** Göran Scharmer, Mats Löfdahl, ISP, SST, Swedish Royal Academy of Sciences; **15mr** Mats Löfdahl, ISP, SST, Royal Swedish Academy of Sciences; **15tl** SOHO(EIT)/NASA/ESA; **15tr** SOHO(EIT)/NASA/ESA; **16br** NASA/ Thierry Legault; **16bl** NASA; **16tr** SOHO(EIT)/NASA/ESA; **16-17** SOHO(EIT)/NASA/ESA; **16l** SST, Royal Swedish Academy of Sciences; **17br** K. Reardon (Osservatorio Astrofisico di Arcetri, INAF) IBIS, DST, NSO; **17mr** NASA / Goddard / SDO AIA Team; **17tr** NASA; **18b** MESSENGER, NASA, JHU APL, CIW; **18br** NASA/ JHU APS/ASU/CIW; Courtesy Science/AAAS; **18l** NASA/Johns Hopkins University Applied Physics Laboratory/Carnegie Institution of Washington; **18mr** MESSENGER, NASA, JHU APL, CIW; **18tr** NASA/Johns Hopkins University Applied Physics Laboratory/Carnegie Institution of Washington; **19b** NASA/ JPL-Caltech; **19t** NASA/JPL-Caltech; **21** NASA; **22b** NASA/ JPL-Caltech; **22mr** NASA/JPL-Caltech; **22tl** Soviet Planetary Exploration Program, NSSDC; **22tr** NASA/JPL/USGS; **23** NASA/ JPL- Caltech, Tim Brown - Pikaia Imaging; **24l** NASA/JPL/USGS; **24–25** NASA/JPL/USGS; **25b** NASA/JPL-Caltech; **25t** NASA/JPL/ USGS; **26t** NASA/JPL-Caltech; **26b** NASA/JPL; **26mc** NASA/JPL; **26tc** NASA/JPL; **26tr** NASA/JPL/USGS; **27l** NASA/JPL/USGS; **27mr** ESA/VIRTIS/INAF-IASF/Obs. de Paris-LESIA/Univ. of Oxford; **27tc** ESA/VIRTIS/INAF-IASF/Obs. de Paris-LESIA/Univ. of Oxford; **27tr** ESA/VIRTIS/INAF-IASF/Obs. de Paris-LESIA/Univ. of Oxford; **29** NASA; **30bl** NASA; **30ml** NASA; **30tl** NASA; **30–31** NASA; **32** NASA Landsat Project Science Office and USGS National Center for EROS; **33b** NASA; **33t** NASA; **34b** NASA/JPL/USGS/; **34t** NASA/JPL/USGS/; **35** STS-129 Crew, NASA; **36b** NASA/JPL/ USGS/; **36t** Giles Sparrow-Pikaia Imaging; **37** Giles Sparrow-Pikaia Imaging; **38m** Robert Gendler; X-ray: NASA/CXC/SAO/J. Drake et al; **38t** NASA; **38m** NASA; **38–39b** Tom Dahl/NASA; **39l** NASA; **39r** NASA/JPL-Caltech; **40bl** NASA; **40tl** NASA; **40tr** NASA; **40-41** I. Mitrofanov et al., LCROSS, LRO, NASA; **41t** NASA; **43** NASA/USGS; **44bl** ESA/DLR/FU Berlin (G. Neukum); **44r** NASA/JPL-Caltech; **44tl** ESA/DLR/FU Berlin (G. Neukum); **45** ESA/DLR/FU Berlin (G. Neukum); **45l** ESA/DLR/FU Berlin (G. Neukum); **45mr** ESA/DLR/FU Berlin (G. Neukum); **45tr** NASA/ JPL-Caltech; **46b** NASA/JPL/Cornell; **46t** USGS Astrogeology Research Program; **47bl** NASA/JPLCaltech/; Cornell; **47br** NASA/ JPL-Caltech/Cornell University; **47ml** NASA/JPL-Caltech/USGS/ Cornell; **47tl** NASA/JPL-Caltech/USGS/Cornell; **48** NASA/JPL/ Malin Space Science Systems; **48–49** NASA/JPL/Malin Space Science Systems; **49bl** ESA/DLR/FU Berlin (G. Neukum); **49br** ESA/DLR/FU Berlin (G. Neukum); **49t** NASA/JPL-Caltech; **50** ESA/DLR/FU Berlin (G. Neukum); **51b** ESA/DLR/FU Berlin (G. Neukum); **51m** ESA/DLR/FU Berlin (G. Neukum); **51t** ESA/DLR/ FU Berlin (G. Neukum); **52l** ESA/DLR/FU Berlin (G. Neukum); **52r** NASA/JPL/University of Arizona; **53l** NASA/JPL/MSSS; **53r** NASA/ JPL-Caltech/University of Arizona; **53t** NASA/JPL-Caltech; **54b** NASA/JPL-Caltech/University of Arizona; **54t** NASA/JPL-Caltech/ University of Arizona; **55** NASA/JPL-Caltech/University of Arizona; **56b** NASA/JPL-Caltech/Cornell; **56t** NASA/JPL-Caltech/Cornell; **57** NASA/JPL-Caltech/Cornell; **58l** NASA/JPL-Caltech/University of Arizona; **58r** NASA/JPL-Caltech/University of Arizona; **59b** NASA/JPL/University of Arizona; **59bm** NASA; **59t** G. Neukum (FU Berlin) et al., Mars Express, DLR, ESA; **59tm** NASA/JPL/Cornell; **60b** NASA/JPL; **60m** NASA/JPL; **60t** NASA/JPL; **61bl** ESA/DLR/ FU Berlin (G. Neukum); **61br** G. Neukum (FU Berlin) et al., Mars Express, DLR, ESA; Acknowledgement: Peter Masek; **61tl** NASA/ JPL-Caltech/University of Arizona/Cornell/Ohio State University; **61tr** HiRISE, MRO, LPL (U. Arizona), NASA; **63** NASA/EAS J. Parker, (Southwest Research Institute), P. Thomas (Cornell Univ.), L. McFadden (Univ. of Marlyland) and M. Mutchler and Z. Levay (STScI); **64b** NASA/JPL; **64tl** NASA/JPL-Caltech; **64tr** NASA/ JPL-Caltech; **65** NASA/JPL-Caltech; **66bl** NASA/JPL-Caltech; **66br** NASA/Goddard Space Flight Center Scientific Visualization Studio; **66ml** NASA/JPL-Caltech; **66mr** NASA/JPL/JHUAPL; **66tc** NASA/JPL/JHUAPL; **66tl** Ben Zellner; (Georgia Southern University), Peter Thomas (Cornell University) and NASA; **66tr** NASA/JPL/JHUAPL; **67b** NASA/JPL/USGS; **67tl** NASA/JPL/USGS; **67tr** INASA/JPL/USGS; **68b** NASA/JPL/USGS; **68t** NASA, ESA, and D. Jewitt (University of California, Los Angeles); **69** ISAS/ JAXA; **69bl** NASA/JPL-Caltech; **69mb** NASA/JPL-Caltech; **69ml** NASA/JPL-Caltech; **69tr** NASA/JPL/University of Arizona; **71** NASA/JPL/Space Science Institute; **72b** NASA/JPL/Space Science Institute; **72l** NASA, ESA, I. de Pater and M. Wong (University of California, Berkeley); **72–73** NASA/JPL-Caltech; **73t,m,b** NASA/ ESA/S. Simon-Miller (Goddard Space Flight Center)/N. Chanover (New Mexico State University/ G. Orton (Jet Propulstion Laboratory); **74br** NASA/JPL-Caltech; **74tl** NASA/JPL; **74tr** NASA/JPL/Space Science Institute; **75l** NASA/JPL/DLR; **75r** NASA/JPLCaltech; **76l** NASA/JPLCaltech; **76–77** NASA/ JPL-Caltech; **77r** NASA/JPL-Caltech; **78l** J. Specer (Lowell Observatory)/NASA; **78rb** NASA/JPL/University of Arizona; **78rt** NASA/JPL/University of Arizona; **79** NASA/JPL; **80m,l,r** NASA/ JPL/DLR; **80tl** NASA/JPL-Caltech; **80tr** NASA/JPL/University of Arizona; **81l** NASA/JPL-Caltech; **81r** NASA/JPL/University of Arizona; **82bl** NASA/JPL/Brown University; **82br** NASA/JPL/ Brown University; **82tl** NASA/JPL-Caltech; **82tr** NASA/JPL/Brown University; **83l** NASA/JPL; **83r** NASA/JPL; **84b** NASA/JPL; **84t** NASA/JPL/DLR; **85** NASA/JPL-Caltech; **86** NASA/JPL-Caltech; **87** NASA/JPL/Space Science Institute; **88b** NASA, ESA, J. Clarke (Boston University), and Z. Levay (STScI); **88t** NASA/JPL/Space Science Institute; **89b** NASA/JPL/Space Science Institute; **89t** NASA/Hubble Heritage Team (STScI/AURA); acknowledgment: R. G./ French (Wellesley College); **90b** Cassini Imaging Team, SSI, JPL, ESA, NASA; **90tl** NASA/JPL/Space Science Institute; **90tr** NASA/JPL/Space Science Institute; **91b** NASA/JPL/Space Science Institute; **91t** NASA/JPL/Space Science Institute; **92bl** NASA/JPL/ Space Science Institute; **92br** NASA/JPL/Space Science Institute; **92tl** Cassini Imaging Team, SSI, JPL, ESA, NASA; Color Composite: Gordan Ugarkovic; **92–93t** VIMS Team, U. Arizona, U. Leicester, JPL, ASI, NASA; **93bl** NASA/JPL/Space Science Institute; **93br** NASA/JPL/Space Science Institute; **93tr** NASA/ JPL/Space Science Institute; **94bl** NASA/JPL/Space Science Institute; **94br** NASA/JPL/Space Science Institute; **94m** NASA/ JPL/Space Science Institute; **94t** NASA/ESA/Hubble Heritage Team (STScI/AURA); **95** ESA/NASA/JPL/University of Arizona; **96tr** NASA/JPL/University of Arizona/University of Nantes; **96b** NASA/JPL; **96tl** NASA/JPL-Caltech/USGS/University of Arizona; **97b** NASA/JPL/USGS; **97br** NASA/JPL/USGS; **97l** NASA/ JPL-Caltech/USGS/University of Arizona; **97tr** ESA/NASA/JPL/ University of Arizona; **98b** NASA/JPL/Space Science Institute; **98tl** NASA/JPL/Space Science Institute; **98tr** NASA/JPL/Space Science Institute; **99** NASA/JPL/Space Science Institute; **100bl** NASA/JPL/Space Science Institute; **100br** NASA/JPL/Space Science Institute; **100tl** NASA/JPL/SSI; Mosaic: Emily Lakdawalla; **100tr** NASA/JPL/Space Science Institute; **101** NASA/JPL/Space Science Institute; **102** NASA/JPL Space Science Institute; **103bl** NASA/JPL/Space Science Institute; **103br** NASA/JPL/Space Science Institute; **103mbr** NASA/JPL/Space Science Institute; **103ml** NASA/JPL/Space Science Institute; **103mr** NASA/JPL/ Space Science Institute; **103tl** NASA/JPL/Space Science Institute; **103tr** NASA/JPL/Space Science Institute; **104b** Cassini Imaging Team, ISS, JPL, ESA, NASA; **104m** NASA/JPL/Space Science Institute; **104tl** NASA/JPL/Space Science Institute/Universities Space Research Association/Lunar & Planetary Institute; **104tr** NASA/JPL/Space Science Institute; **105** NASA/JPL/Space Science Institute; **106b** NASA/JPL/Space Science Institute; **106m** Cassini Imaging Team, ISS, JPL, ESA, NASA; **106t** Cassini Imaging Team, ISS, JPL, ESA, NASA; **106–107** NASA/JPL/Space Science Institute; **107b** NASA/JPL/Space Science Institute; **107bm** NASA/ JPL/Space Science Institute; **107m** Cassini Imaging Team, SSI, JPL, ESA, NASA; **107t** NASA/JPL/Space Science Institute; **107tm** NASA/JPL/Space Science Institute; **108br** Erich Karoschka (University of Arizona) and NASA; **108tl** NASA/JPL-Caltech; **109br** NASA/JPL; **109l** NASA/JPL-Caltech; **109mr** NASA/JPL; **109tr** NASA/JPL-Caltech; **110b** NASA/JPL; **110tl** NASA/ESA/M. Showalter (SETI Institute); **110tr** NASA/JPL-Caltech; **111br** NASA/JPL; **111l** NASA/JPL-Caltech; **111mr** NASA/JPL; **112bl** NASA/JPL-Caltech; **112r** NASA/JPLCaltech; **112t,b** Tim Brown – Pikaia Imaging; **113br** NASA/JPL; **113mr** NASA/JPL/ Universities Space Research Association/Lunar & Planetary Institute; **113tl** NASA/JPL; **113tr** NASA/JPL; **115** NASA, ESA, H. Weaver; (JHU/APL), A. Stern (SwRI), and the HST Pluto Companion Search Team; **116bl** NASA, ESA and M. Brown (Caltech); **116br** NASA/ESA/M. Buie (Southwest Research Institute); **116tl** NASA, ESA, and M. Brown (California Institute of; Technology); **116tr** Dr. R. Albrecht, ESA/ESO Space Telescope European Coordinating Facility; **117** M. Brown/Caltech; **118** NASA, NOAO,; NSF, T. Rector (University of Alaska Anchorage), Z. Levay and L.Frattare (Space Telescope Science Institute); **119b** ESA; **119t** Halley Multicolor Camera Team, Giotto Project, ESA; **120t** NASA, ESA, H. Weaver (APL/JHU), M. Mutchler and Z. Levay (STScI); **120–121** NASA/JPL-Caltech ; **122b** NASA; **122t** Sebastian Deiries (ESO); **123** Canada-France-Hawaii Telescope/ Coelum/J.-C. Cuillandre & G. Anselmi; **124bc** NASA/JPL/ CalTech; **124bl** NASA/JPL-Caltech/UMD; **124bml** NASA/ JPL-Caltech/UMD; **124br** NASA/JPL; **124tl** NASA/JPL-Caltech; **124tml** NASA/JPL/UMD; **124tr** NASA/JPL; **125bl** NASA/ JPL-Caltech/UMD/Brown; **125t** BNASA/yron Bergert; **126bl** NASA/JPL; **126ml** NASA/JPL; **126r** NASA/JPL; **126tl** NASA/JPL; **127br** NASA/JPL; **127bl** NASA/JPL/STScI; **127t** SonotaCo Network, Japan; **128** Harvard College Observatory/Science Photo Library; **129** NASA/JPL-Caltech/University of Minnesota; **132–133** NASA, ESA, P. Kalas and J. Graham (University of California, Berkeley), and M. Clampin (NASA's Goddard Space Center); **135** NASA/CXC/SAO; **136** A. Fujii; **137b** ESO/R.-D.Scholz et al.; **137t** NASA, H.E. Bond and E. Nelan (Space Telescope Science Institute, Baltimore, Md.); M. Barstow and M. Burleigh (University of Leicester, U.K.); and J.B. Holberg (University of Arizona); **138t** NASA, ESA and D. Golimowski; **138b** NASA/JPL-Caltech/UCSC; **138l** National Research Council Canada; **138m** NASA and F.M. Walter; (State University of New York at Stony Brook); **139br** NASA; **139l** NASA; **139tr** NASA; **140** NASA/JPL-Caltech/J. Stauffer (SSC/Caltech); **141b,l,r** John Davis; **141t** NASA, ESA and AURA/Caltech; **143** Image Data: Digitized Sky Survey; Color Composite: Noel Carboni; **144** NASA JPL-Caltech, Harvard-Smithsonian; **145** Robert Gendler, Jim Misti, Steve Mazin; **146b** NASA/CXC/SAO/M. Karovska et al; **146m** NASA/JPL-Caltech/ POSS-II/DSS; **146t** NASA/JPL-Caltech/POSS-II/DSS; **147** Robert Gendler, Martin Pugh; **148b** European Southern Observatory; **148m** NASA, John Krist (Space Telescope Science Institute), Karl Stapelfeldt (Jet Propulsion Laboratory), Jeff Hester (Arizona State University), Chris Burrows (European Space Agency/Space Telescope Science Institute); **148t** NASA/JPL-Caltech/UCSC; **149tc** Andrea Dupree (Harvard-Smithsonian CfA), Ronald Gilliland (STScI), NASA and ESA; **149bl** Pierre Kervella, NaCo, VLT, ESO; **149l** A. Fujii; **149tr** Andrea Dupree (Harvard-Smithsonian CfA), Ronald Gilliland (STScI), NASA and ESA; **150ml** C. Robert O'Dell and Kerry P. Handron (Rice University)/NASA; **150tl** NASA/NOAO/ ESA/Hubble Helix Nebula Team/M. Meixner (STScI)/T. A. Rector (NRAO); **150tr** NASA, NOAO, ESA, the Hubble Helix Nebula; Team, M. Meixner (STScI), and T.A. Rector (NRAO); **151** NASA/NOAO/ ESA/Hubble Helix Nebula Team/M. Meixner (STScI)/T. A. Rector (NRAO); **152bl** NASA/JPL-Caltech/UCLA/DSS; **152br** NASA/ JPL-Caltech/UCLA; **152t** NASA/JPL-Caltech/Harvard-Smithsonian CfA; **153b** ESA/PACS & SPIRE Consortium, Amélia Stutz, Max-Planck-Institut für Astronomie; **153t,l,r** NASA/ JPL-Caltech/Harvard-Smithsonian; **154b** European Southern Observatory; **154m** Canada-France-Hawaii Telescope/ Coelum/J.-C. Cuillandre & G. Anselmi; **154t** Robert Gendler and Jim Misti; **155** Image Data: Digitized Sky Survey; Color Composite: Noel Carboni; **156l** Digitized Sky Survey, ESA/ESO/ NASA FITS Liberator; **156–157** ESO and Digitized Sky Survey 2/ Davide De Martin; **157b** NASA and The Hubble Heritage Team (STScI/; AURA); **157t** NASA; **159** NASA/Stephen Leshin; **160l** NASA/CXC/PSU/G. Pavlov et al.**160** NASA/CXC/PSU/G.Pavlov et al; **160–161** Allan Cook & Adam Block, NOAO, AURA, NSF; **161** NASA/Alessandro Falesiedi; **162** NASA/JPL/STScI; **162–163** European Southern Observatory; **163** Daniel López, IAC; **164–165** NASA, C.R. O'Dell and S.K. Wong (Rice University); **165** NASA; K.L. Luhman (Harvard-Smithsonian Center for Astrophysics,

Cambridge, Mass.); and G. Schneider, E. Young, G. Rieke, A. Cotera, H. Chen, M. Rieke, R. Thompson (Steward Observatory, University of Arizona, Tucson, Ariz.); **166** ESA and the SPIRE & PACS consortia, P. André (CEA Saclay) for the Gould's Belt Key Programme Consortia; **167** Adam Block/NOAO/AURA/NSF; **168–169** NASA/JPL-Caltech/ Laboratorio de Astrofísica Espacial y Física Fundamental; **170** NASA,ESA, M. Robberto (Space Telescope Science Institute/ESA) and; the Hubble Space Telescope Orion Treasury Project Team; **171r** NASA/ JPL-Caltech/J. Stauffer (SSC/Caltech); **171** NASA/ESA/M. Robberto (Space Telescope Science Institute/ESA)/Hubble Space Telescope Orion Treasury Project Team; **172** Infrared Processing and Analysis Center, Caltech/JPL; **172–173** AAAS/Science (ESA XMM-Newton and NASA's Spitzer Space Telescope); **173** C.R. O'Dell/Rise University NASA/ESA; **174** NASA, ESA, M. Robberto (STScI/ESA), the HST Orion Treasury Project Team, & L. Ricci (ESO); **175b** NASA, J. Bally (University of Colorado), H. Throop (SWRI), C. R. O'Dell (Vanderbilt University); **175t** Mark McCaughrean (Max-Planck-Institute for Astronomy), C. Robert O'Dell (Rise University), and NASA; **176** NASA, The Hubble Heritage Team (AURA/STScI); **177b** NASA and The Hubble Heritage Team (STScI/AURA); **177t** Robert Gendler, Jan-Erik Ovaldsen; **178l** European Southern Observatory; **178–179** T. A. Rector and B. A. Wolpa /NOAO/AURA/NSF; **180–181** ESO/J. Emerson/VISTA. Acknowledgment: Cambridge Astronomical Survey Unit; **181b** NASA/NOAO/ESA/Hubble Heritage Team (STScI/AURA); **181t** T.A.Rector (NOAO/AURA/NSF); and Hubble Heritage Team (STScI/AURA/NASA); **182** University of Colorado, University of Hawaii and NOAO/AURA/NSF; **183bl** Hubble Heritage Team (STScI/AURA/NASA); **183br** NASA and The Hubble Heritage Team (STScI/AURA); **183ml** NASA and The Hubble Heritage Team (STScI/AURA); **183mr** NASA/Robert Nemiroff (MTU) & Jerry Bonnell (USRA); **183t** T. A. Rector/University of Alaska Anchorage, H. Schweiker/WIYN and NOAO/AURA/NSF; **184**bc Bruce Balick; (University of Washington), Jason Alexander (University of Washington), Arsen Hajian (U.S. Naval Observatory), Yervant Terzian (Cornell University), Mario Perinotto (University of; Florence, Italy), Patrizio Patriarchi (Arcetri Observatory, Italy) and NASA; **184bl** Bruce Balick (University of Washington), Vincent Icke (Leiden University, The Netherlands), Garrelt Mellema (Stockholm University), and NASA; **184br** NASA; ESA; Hans Van Winckel (Catholic Univeristy of Leuven, Belgium); and Martin Cohen (University of California, Berkeley); **184tl** NASA/Hubblesite; **184tr** Hubble Heritage Team (AURA/; STScI/NASA); **185** Davide De Martin (Skyfactory); **186b** T.A. Rector (University of Alaska Anchorage) and WIYN/NOAO/AURA/NSF; **186t** NASA/Rolf Geissinger; **187bl** NASA/ Thomas V. Davis (tvdavisastropix.com); **187br** NASA and The Hubble Heritage Team (AURA/STScI); **187t** NASA/JPL/Caltech; **188l** T. A. Rector (NRAO/AUI/NSF and NOAO/ AURA/NSF) and B. A. Wolpa (NOAO/AURA/NSF); **188–189** NASA, ESA, the Hubble Heritage Team (STScI/AURA)-ESA/Hubble Collaboration, and the Digitized Sky Survey 2, acknowledgment: J. Hester (Arizona State University) and Davide De Martin (ESA/ Hubble); **189b** T. Rector/Univerity of Alaska Anchorage and WIYN/ NOAO/AURA/NSF; **189t** NASA/ESA/Hubble Heritage (STScI/ AURA)-ESA/Hubble Collaboration; **190–191** NASA/JPL-Caltech/ University of Arizona/Texas A&M University; **193** NASA/ JPL-Caltech/L. Allen (Harvard-Smithsonian CfA) & Gould's Belt Legacy Team; **194b** NASA, Andrew Fruchter and; the ERO Team (Sylvia Baggett (STScI), Richard Hook (ST-ECF), Zoltan Levay (STScI)); **194ml** NASA/Hubble Heritage Team (STScI/AURA); **194mr** NASA, ESA and The Hubble Heritage Team (STScI/AURA); **194tl** NASA/JPL-Caltech/UCLA; **194tr** ESA, Hubble, R. Sahai (JPL), NASA; **195** Canada-France-Hawaii Telescope/Coelum/J.-C. Cuillandre & G. Anselmi; **196b** NASA/Tony Hallas; **196tl** NASA, ESA, HEIC, and The Hubble; Heritage Team (STScI/AURA); **196tr** J. P. Harrington (U. Maryland) & K. J. Borkowski (NCSU) HST, NASA; **197** A. Zijlstra (UMIST) et al., ESA, NASA; **198b** NASA/ESA/ Hubble Heritage Team (STScI/AURA); **198m** ESA/PACS/SPIRE/ HOBYS Consortia; **198tl** NASA/ESA/K. Noll (STScI); **198tr** NASA/ ESA/Hubble SM4 ERO Team; **199b** Garrelt Mellema (Leiden University) et al., HST, ESA, NASA; **199t** NASA and The Hubble Heritage Team (AURA/STScI); **200m** NASA/ESA and R. Sahai/JPL; **200tl** NASA, JPL-Caltech, WISE Team; **200–201** Steve Mazlin, Jack Harvey, Rick Gilbert, and Daniel Verschatse; (Star Shadows Remote Observatory, PROMPT, CTIO); **201tr** ESO; **201br** NASA, H. Ford (JHU), G. Illingworth (UCSC/LO), M. Clampin (STScI), the ACS Science Team, and ESA; **202b** ESA and the PACS, SPIRE & HSC consortia, F. Motte (AIM Saclay,CEA/IRFU-CNRS/INSU-U. ParisDidedrot) for the HOBYS key programme; **202tl** NASA/ Robert Gendler; **202tr** NASA/JPL-Caltech/Univ. of Ariz.; **203** Nick Wright (University College London), IPHAS Collaboration; **204bl** ESA/NASA & Valentin Bujarrabal (Observatorio Astronomico Nacional, Spain); **204br** X-ray: NASA/UIUC/Y. Chu & R. Gruendl et al. Optical: SDSU/MLO/Y. Chu et al.; **204t** R. Sahai and J. Trauger (JPL), NASA/ESA; **205** T. A. Rector/University of Alaska Anchorage, T. Abbott and NOAO/AURA/NSF ; **206t** Hubble Heritage Team (AURA/STScI/NASA); **206b** NASA/Swift/Stefan Immler and Eric Grand; **207** NASA,; H. Ford (JHU), G. Illingworth (UCSC/LO), M.Clampin (STScI), G. Hartig (STScI), the ACS Science Team, and ESA; **208** I.A. Rector/University of Alaska Anchorage, T. Abbott and; NOAO/AURA/NSF; **209** NASA, ESA, and the Digitized Sky Survery 2, acknowledgment: Davide De Martin (ESA/Hubble); **210b** NASA/JPL-Caltech; **210m** NASA/JPL-Caltech/A. Marston (ESTEC/ESA) & A. Noriega-Crespo (SSC/Caltech); **210t** NASA and The Hubble Heritage Team (STScI/AURA) Acknowledgment: Bo Reipurth (University of Hawaii); **210–211** NASA/JPL-Caltech/ UCLA; **211b** NASA/JPL-Caltech/UCLA; **211t** NASA/JPL-Caltech/ UCLA; **212l** NASA/CXC/Rutgers/J.Hughes et al; **212r** NASA/CXC/ ASU/J. Hester et al.; **213** NASA, ESA, and the Hubble Heritage (STScI/AURA)-ESA/Hubble Collaboration, acknowledgment: J. Maiz Apellaniz (Institute of Astrophysics of Andalucia, Spain); **214l** NASA, The Hubble Heritage Team (STScI/; AURA); **214r** NASA, The Hubble Heritage Team (AURA/STScI); **215bl** NASA, ESA, Zolt Levay (STScI); **215br** NOAO/AURA/NSF; **215tl** NASA/JPL-Caltech/ Harvard-Smithsonian CfA; **215tr** NASA/CXC/Rutgers/J.Warren & J.Hughes et al; **216l** NASA and the Hubble Heritage Team (AURA/ STScI); p120 NASA, ESA, and The Hubble Heritage Team (AURA/ STScI); **216–217** NASA, ESA, and The Hubble Heritage Team (AURA/STScI); **219** NASA, Jess Hester and Paul Scowen Arizona State University; **220** T.A.Rector (NRAO/AUI/NSF and NOAO/ AURA/NSF) and B.A.Wolpa (NOAO/AURA/NSF); **221** NASA, ESA, STScI, J. Hester and P. Scowen (Arizona State University); **222b** INASA, ESA, N. Smith (U. California, Berkeley) et al., and The Hubble Heritage Team (STScI/AURA); **222mt** NASA, ESA, N. Smith (U. California, Berkeley) et al., and The Hubble Heritage Team (STScI/AURA); **222tl** Nathan Smith, University of Minnesota/ NOAO/AURA/NSF; **222tr** NASA, ESA, N. Smith (University of California, Berkeley); and The Hubble Heritage Team (STScI/ AURA); **223** NASA/ESA M. Livio and the Hubble Team; **224** ESO/J. Emerson/VISTA. Acknowledgment: Cambridge Astronomical Survey Unit; **225l** ESO; **225r** Credit for Hubble Image: NASA, ESA, N. Smith (University of California, Berkeley), and The Hubble Heritage Team (STScI/AURA), Credit for CTIO Image: N. Smith (University of California, Berkeley)and NOAO/AURA/NSF; **226** NASA/ Adam Block, Mt. Lemmon SkyCenter, U. Arizona; **227br** NASA, ESA, and the Hubble Heritage Team (AURA/STScI); **227tr** NASA/JPL-Caltech/NOAO; **228** NASA, ESA and Jesus Maiz Apellaniz (Institute of Astrophysics of Andalucia, Spain); **229b** Raghvendra Sahai and John Trauger (JPL), the WFPC**2** science team, and NASA; **229t** Davide De Martin (ESA/Hubble), the ESA/ ESO/ NASA, Photoshop FITS Liberator & Digitized Sky Survey 2; **231** ASA, ESA and Jesoes Maz Apellÿniz (Instituto de astrofisica de Andalucia, Spain). Acknowledgement: Davide De Martin (ESA/ Hubble); **230** ESO/IDA/Danish 1.5 m/ R. Gendler, U.G. Jørgensen, J. Skottfelt, K. Harpsøe; **232** NASA, The Hubble Heritage Team (AURA/STScI); **233b** NASA/ESA/N. Smith (University of California, Berkeley)/The Hubble Heritage Team (STScI/AURA)/NOAO/AURA/ NSF; **233m** NASA/ESA/Hubble Heritage Team (STScI/AURA); **233t** AURA/NOAO/NSF; **234b** NASA, ESA, and the Hubble Heritage (STScI/AURA)-ESA/Hubble Collaboration, acknowledgment: R. Fesen (Dartmouth College) and J. Long (ESA/Hubble); **234r** NASA / JPL-Caltech / O. Krause (Steward Observatory); **234t** NASA, The Hubble Heritage Team (AURA/STScI); **235** NASA/JPL-Caltech/o. Krause (Steward Observatory); **236l** NASA, The Hubble Heritage Team (AURA/STScI); **236r** Jean-Charles Cuillandre (CFHT), Hawaiian Starlight, CFHT; **237br** P. Garnavich (CfA); **237l** ESO/ NASA/JPL-Caltech/S. Kraus; **237**mc F. Paresce, R. Jedrzejski (STScI), NASA/ESA; **237mr** F. Paresce, R. Jedrzejski (STScI), NASA/ESA; **238bl** Hubble Heritage Team (AURA/STScI/NASA); **238br** NASA, Donald Walter (South Carolina State University), Paul Scowen and Brian Moore (Arizona State University); **238t** NASA/JPL-Caltech/S. Carey (SSC/Caltech); **239b** NASA/ESA/ JHU/R.Sankrit & W.Blair; **239t** NASA, ESA and L. Bedin; **240** NASA/JPL-Caltech/University of; Wisconsin; **241l** ESA/Hubble and NASA; **241r** Canada-France-Hawaii Telescope/J.-C. Cuillandre; **242bl** NASA, ESA, J. Anderson and R. van der Marel (STScI); **242br** NASA, The Hubble Heritage Team (AURA/STScI); **242tl** NOAO/AURA/NSF; **242tr** NASA, ESA, and The Hubble Heritage Team (STScI/AURA); **243** NASA, ESA, M. Livio and the Hubble 20th Anniversary Team (STScI); **243b** (Left): Very Large Telescope/European Southern Observatory, R. Kotak and H. Boffin (ESO). (Right): NASA, ESA, and G. Meylan (Ecole Polytechnique Federale de Lausanne); **243t** NASA, DOE, Fermi LAT Collaboration; **244l** NASA/McGill/V. Kaspi et al; **244r** NASA; **245b** NASA/CXC/CfA/P. Slane et al.; **245m** NASA/CXC/CfA/P. Slane et al.; **245t** T. Rector (U. Alaska Anchorage), Gemini Obs., AURA, NSF; **246l** ESA; **246–247** Wolfgang Brandner; (JPL/IPAC), Eva K. Grebel (Univ. Washington), You-Hua Chu (Univ. Illinois Urbana-Champaign), and NASA; **247** NASA and The Hubble Heritage Team (AURA/STScI); **248b** NASA/ESA/Hubble Heritage Team (STScI/AURA); acknowledgment: C. Bailyn (Yale University), W. Lewin (Massachusetts Institute of Technology), A. Sarajedni (University of Florida), and W. van Altena (Yale University); **248t** X-ray (NASA/CXC/NCSU/S.Reynolds et al.); **249b** NASA, ESA, Hubble Space Telescope; **249t** NASA,; Don Figer, STScI; **251** Atlas Image courtesy of 2MASS/Umass/IPAC-Caltech/NASA/NSF/G. Kopan, R. Hurt; **252t** NASA, ESA, SSC, CXC, and STScI; **252b** NASA/UMass/D. Wang et al; **253bl** NASA/CXC/Caltech/M. Muno et al; **253br** ESO; **253t** NASA/JPL-Caltech/S. Stolovy (SSC/ Caltech); **254b** NASA/UMass/D. Wang et al; **254t** NRAO/VLA F. Zadeh et al; **255b** ESA and the HFI Consortium.; **255t** Image composite by Ingrid Kallick of Possible Designs, Madison Wisconsin; **261** NASA, ESA, and M. Davis (University of California, Berkeley); **262** A. Nota (ESA/STScI) et al., ESA, NASA; **263** NASA/ JPL-Caltech/M. Meixner (STScI) and the SAGE Legacy Team; **264** NASA/ESA/M. Livio (STScI); **265b** NASA, ESA; **265tl** NASA/CXC/ STScI/JPL-Caltech/UIUC/Univ. of Minn.; **265tr** NASA and The; Hubble Heritage Team (STScI/AURA), A. Cool (San Fransisco State University and J. Anderson (STScI); **266b** NASA/CXC/ Rutgers/J.Hughes et al; **266tl** NASA, ESA; & Mohammad Heydari-Malayeri (Observatoire de Paris, France); **266–267** NASA, ESA, Mohammad Heydari-Malayeri (Observatoire; de Paris, France); **267** Hubble Heritage Team (AURA/STScI/NASA); **268** ESA/NASA. ESO and Danny LaCrue; **269b** N. Walborn (STScI) et al., WFPC**2**, HST, NASA; **269t** N. Walborn (STScI) et al., WFPC2, HST, NASA; **270** NASA, ESA, and Martino Romaniello (European Southern Observatory, Germany); **270–271** The Hubble Heritage Team (AURA/STScI/NASA); **271b** NASA, ESA, Y. Nazé (University of Liège, Belgium) and Y.-H. Chu (University of Illinois, Urbana); **271t** NASA, ESA, K. France (University of Colordo, Boulder), and P. Challis and R. Kirshner (Harvard-Smithsonian Center for Astrophysics); **272b** NASA, ESA, Y. Nazé (University of Liège, Belgium) and Y. H. Chu; (University of Illinois, Urbana); **272t** NASA, The Hubble Heritage Team (AURA/STScI); **273** Gemini Observatory/AURA; **274b** NASA/CXC/SOA; **274l** NASA/ESA/HEIC/ Hubble Heritage Team (STScI/AURA); **274t,r,c** NASA/CXC/SAO; **275b** NASA/JPL-Caltech/K. Gordon (STScI); **275t** ESA/Hubble and Digitized Sky Survey; **276–277** NASA, ESA and The Hubble Heritage Team (STScI/AURA) -ESA/Hubble Collaboration; **279** NASA, ESA, and The Hubble Heritage Team (STScI/AURA); **280** NASA; **281** NASA; **282–283** T.A. Rector and B.A. Wolpa/NOAO/ AURA/NSF); **284bl** NASA and The Hubble Heritage Team (STScI/ AURA); **284tl** Tim Brown-Pikaia Imaging; **285** NASA; **286bl** NASA, ESA and T.M. Brown (STScI); **286bml** NASA/JPL-Caltech/K. Gordon (University of Arizona); **286tl** NASA/JPL-Caltech/K. Gordon (University of Arizona); **286tml** NASA/JPL-Caltech/P. Barmby (Harvard-Smithsonian CfA); **286–287** infrared: ESA/ Herschel/PACS/SPIRE/J. Fritz, U. Gent; X-ray: ESA/XMM-Newton/ EPIC/W. Pietsch, MPE; **287br** X-ray (NASA/CXC/SAO/Li et al.), Optical (DSS); **287tr** NASA, ESA, and M. Davis (University of California, Berkeley); **288bl** NRAO/AUI; **288br** NRAO/AUI; **288tl** NRAO/AUI and NOAO/AURA/NSF; **288–289** T. A. Rector (NRAO/ AUI/NSF and NOAO/AURA/NSF); **289r** NASA and The Hubble Heritage Team (AURA/STScI); **290l** NASA, ESA, J. Dalcanton and B. Williams (University of Washington); **290r** M. Schirmer (IAEF, Bonn), W. Gieren (Univ. de Concepción, Chile), et al., ESO; **291b** NASA/JPL-Caltech/J. Turner (UCLA); **291tl** NASA/CXC/UCSB/C.

Martin et al.; **291tr** ESA, NASA and P. Anders (Göttingen University Galaxy Evolution Group, Germany); **292** E.J. Schreier (STScI) and NASA; **293b** NASA, The Hubble Heritage Team (AURA/STScI); **293t** Jack O. Burns (University of Missouri)/David Clarke (St. Mary's University, Nova Scotia); **294** NASA, ESA, The Hubble; Heritage Team (STScI/AURA) J. Gallagher (University of Wisconsin), M. Mountain (STScI), and P. Puxley (National Science Foundation; **295b** NASA/JPL-Caltech/S. Willner (Harvard-Smithsonian Center for Astrophysics); **295t** NASA/ESA/Hubble Heritage Team (STScI/AURA); **296t** NASA/ESA/A. Aloisi (STScI/ESA)/Hubble Heritage (STScI/AURA)-ESA/Hubble Collaboration; **296b** NASA, Andrew S. Wilson (University of Maryland); Patrick L. Shopbell (Caltech); Chris Simpson (Subaru Telescope); Thaisa Storchi-Bergmann and F. K. B. Barbosa (UFRGS, Brazil); and Martin J. Ward (University of Leicester, UK); **297** ESO; **298tl** NASA/JPL-Caltech/UCLA; **298–299** A. Watson (UNAM) et al., Hubble Heritage Team (STScI / AURA), NASA; **299b** Henri Boffin (ESO), FORS1, 8.2-meter VLT, ESO; **299t** NASA, ESA, Anne Pellerin (STScI); **300** NASA/CXC/U. Leicester/U. London/R. Sorai/K. Wu; **301bl** NASA/JPL-Caltech/VLA/MPIA; **301br** NASA/ESA/Hubble Heritage Team (STScI/AURA); **301t** European Southern Observatory; **302l** Acquisition-Martin Winder, Processing-Warren Keller; **302–303** NASA and The Hubble Heritage Team (AURA/; STScI); **303b** Johannes Schedler (Panther Observatory); **303t** NASA/JPL-Caltech; **304–305t** Hubble Legacy Archive, ESA, NASA; Processing: Nikolaus Sulzenauer; **304bl** Bernie and Jay Slotnick, Adam Block, AOP, NOAO, AURA, NSF; **304br** X-ray: NASA/CXC/Univ. of Maryland/A. S. Wilson et al; Optical: Pal. Obs. DSS; IR: NASA/JPL-Caltech; VLA: NRAO/AUI/NSF; **305b** NASA and the Hubble Heritage Team (STScI/AURA); **306t** NASA/ESA/STScI/JHU/K. Kuntz et al; IR: NASA/JPL-Caltech/STScI/K. Gordon; **306b** NASA/CXC/K. Kuntz (JHU); **307** NASA and ESA, Acknowledgment: K. D. Kuntz (GSFC), F. Bresolin (University of Hawaii), J. Trauger (JPL), J. Mould (NOAO), and Y. -H. Chu (University of Illinois, Urbana); **308bl** NASA/UMass/Q. D. Wang et al; **308ml** NASA, ESA, and the Hubble Heritage Team (STScI/AURA); **308tl** NASA and The Hubble Heritage Team (STScI/AURA); **308–309** NASA/ESA/Hubble Heritage (STScI/AURA)-ESA/Hubble Collaboration; **309tr** NASA/JPL-Caltech/B. E. K. Sugerman (STScI); **310bl** Hubble Legacy Archive, ESA, NASA; **310tl** NASA/JPL-Caltech/CTIO; **310–311** NASA, ESA, S. Beckwith (STScI), and The Hubble Heritage Team (STScI/AURA); **311br** NASA, ESA, S. Beckwith (STScI), and The Hubble Heritage Team (STScI/AURA); **311tr** NASA, ESA, M. Regan and B. Whitmore (STScI), R. Chandar (University of Toledo), S. Beckwith (STScI), and the Hubble Heritage Team (STScI/AURA); **312b** NASA, ESA, R.M. Crockett (University of Oxford, U.K.), S. Kaviraj (Imperial College London and University of Oxford, U.K.), J. Silk (University of Oxford), M. Mutchler (Space Telescope Science Institute, Baltimore), R. O'Connell (University of Virginia, Charlottesville), and the WFC3 Scientific Oversight Committee; **312t** Ken Crawford (Rancho Del Sol Obs.), Collaboration: David Martinez-Delgado (MPIA, IAC), et al.; **3131b** NASA/ESA; **313l** NASA/ESA/Hubble Heritage Team (STScI/AURA); **315** NASA, Gerald Cecil (University of North Carolina), Sylvain Veilleux (University of Maryland), Joss Bland-Hawthorn (Anglo-Australian; Observatory), and Alex Filippenko (University of California at Berkeley); **316bl** Canada-France-Hawaii Telescope/Coelum/J.-C. Cuillandre & G. Anselmi; **316br** NASA/CXC/CfA/W. Forman et al.; Radio: NRAO/AUI/NSF/W. Cotton; Optical: NASA/ESA/Hubble Heritage Team (STScI/AURA), and R. Gendler; **316tl** NOAO/AURA/NSF; **317t** NASA, ESA, and J. Madrid (McMaster University); **317b** X-ray: NASA/CXC/KIPAC/N. Werner et al Radio: NSF/NRAO/AUI/W. Cotton; **318bl** NASA, JPL-Caltech, SINGS Team (SSC); **318br** SSRO/PROMPT and NOAO/AURA/NSF; **318t** R. Jay GaBany (Cosmotography.com); **319b** NASA, ESA, and Z. Levay (STScI); **319t** Piotrek Sadowski; **320** NASA and Jeffrey Kenney (Yale University); **321** NASA, ESA, The Hubble Key Project Team, and The High-Z Supernova Search Team; **322b** Gemini Observatory/AURA; **322t** NASA, The Hubble Heritage Team (AURA/STScI); **323bl** & R B. Woodgate (GSFC), G. Bower (NOAO) and NASA; **323tl** Hubble Heritage Team (AURA/STScI/NASA); **323tr** X-ray (NASA/CXC/MPE/A.Finoguenov et al.); Radio (NSF/NRAO/VLA/ESO/R.A.Laing et al); Optical (SDSS); **324l** ESA, Hubble, NASA; **324–325** NASA, ESA, and B. Whitmore (STScI); **325br** Brad Whitmore (STScI) and NASA; **325tr** NASA/CXC/SAO/G. Fabbiano et al; **326** NASA, ESA, and the Hubble Heritage Team (STScI/AURA), Acknowledgment: P. Knezek (WIYN); **327bl** NASA, ESA, and The Hubble Heritage Team (STScI/AURA), J. Gallagher (University of Wisconsin), M. Mountain (STScI), and P. Puxley (National Science Foundation); **327br** NASA/JPL-Caltech/R. Kennicutt (University of Arizona) and the SINGS Team; **327tr** Hubble Heritage Team (AURA/ STScI/ NASA); **328bl** NASA, The Hubble Heritage Team (AURA/STScI); **329** NASA; **330** NASA/JPL-Caltech/SSC; **331b** NASA/Swift Science Team/Stefan Immler.; **331t** H. Boffin, H. Heyer, E.Janssen (ESO), FORS2, European Southern Observatory; **332b** J. Trauger, JPL and NASA; **332tr** ESO/IDA/Danish 1.5 m/R. Gendler, J.E. Ovaldsen, C. C. Thöne and C. Féron; **333b** Jeffrey Newman (UC Berkeley), NASA; **333t** NASA, The Hubble; Heritage Team and A. Riess (STScI); **334** NASA, ESA and W. Harris (McMaster University, Ontario, Canada); **335br** NASA and The Hubble Heritage Team (STScI/AURA); **335tr** L. Ferrarese (Johns Hopkins University) and NASA; **336** NASA/JPL-Caltech/STScI/Vassar; **337** NASA, ESA, and the Hubble Heritage Team (STScI); **338bl** NASA and The Hubble Heritage Team (STScI/AURA); **338ml** NASA/JPL-Caltech/P. N. Appleton (SSC/Caltech); **338–339** NASA and The Hubble Heritage Team (STScI/AURA); **339b** X-ray: NASA/CXC/KIPAC/s. Allen et al; Radio: NRAO/VLA/ G. Taylor; Infrared: NASA/ESA/McMaster Univ./W. Harris; **339t** NASA and The Hubble Heritage Team (STScI/AURA); **340b** NASA, J. English (U. Manitoba), S. Hunsberger, S. Zonak, J. Charlton, S. Gallagher (PSU), and L. Frattare (STScI); **340t** NASA, ESA, J. English (U. Manitoba), and the Hubble Heritage Team (STScI/AURA); Acknowledgement: S. Gallagher (U. Western Ontario); **341l** NASA and The Hubble Heritage Team (STScI/AURA), Acknowledgment: Roger Lynds (KPNO/NOAO); **341r** ESO; **342** NASA and The Hubble; Heritage Team (STScI/AURA); **343bl** NASA/DOE/Fermi LAT Collaboration; **343br** C. Conselice (Caltech), WIYN, AURA, NOAO, NSF; **343t** NASA, ESA, Hubble Heritage (STScI/AURA); A. Fabian (IoA, Cambridge U.), L. Frattare (STScI), CXC, G. Taylor, NRAO,VLA; **344l** 2MASS/T. H. Jarrett, J. Carpenter, & R. Hurt; **345bl** NASA/JPL-Caltech/Max Planck Institute; **345br** NASA/JPL-Caltech/Max-Planck Institute/P. Appleton (SSC/Caltech); **345tr** NASA/Dietmar Hager; **344–345** NASA/Jayanne English (University of Manitoba)/Sally Hunsberger (Pennsylvania State University)/Zolt Levay (Space Telescope Science Institute)/Sarah Gallagher (Pennsylvania State University)/Jane Charlton (Pennsylvania State University); **346–347** NASA,; ESA, and The Hubble Heritage Team (AURA/STScI); **348bl** NASA/ESA/Hubble Heritage (STScI/AURA)-ESA/Hubble Collaboration/A. Evans (University of Virginia, Charlottesville/NRAO/Stony Brook University); **348br** NASA, ESA, and the Hubble Heritage Team (STScI/AURA); **348t** NASA, William C. Keel; (University of Alabama, Tuscaloosa); **349** NASA, H. Ford (JHU), G. Illingworth (UCSC/LO), M.Clampin (STScI); G. Hartig (STScI), the ACS Science Team, and ESA; **350** NASA, ESA and the Hubble Heritage (STScI/AURA)-ESA/Hubble Collaboration. Acknowledgment: M. West (ESO, Chile); **351b** X-ray: NASA/CXC/MIT/C.Canizares, M.Nowak; Optical: NASA/STScI; **351tl** NASA/CXC/AIfA/D. Hudson & T. Reiprich et al; **351tr** NASA/CXC/MPE/S. Komossa et al; **352bl** X-ray: NASA/CXC/Penn State/G. Garmire; Optical: NASA/ESA/STScI/M. West; **352br** NASA, ESA, and the Hubble Heritage Team (STScI/AURA); Acknowledgment: K. Cook (LLNL) et al.; **352tl** NASA, ESA, and The Hubble Heritage (STScI/AURA)-ESA/Hubble Collaboration, acknowledgment: M. West (ESO, Chile); **352tr** NASA/JPL/Caltech/L. Jenkins (GSFC); **353** NASA, ESA, K. Cook (LLNL); **354b** NASA, ESA, and M. Livio (STScI); **354t** The Hubble Heritage Team (AURA/STScI/NASA); **355br** NASA/JPL-Caltech/STScI; **355l** Canada-France-Hawaii Telescope/Coelum/J.C. Cuillandre & G. Anselmi; **355tr** ASA, ESA, the Hubble Heritage Team (STScI/AURA)-ESA/Hubble Collaboration and A. Evans (University of Virginia, Charlottesville/NRAO/Stony Brook University); **356l** NASA, ESA, the Hubble Heritage Team (STScI/AURA)-ESA/Hubble Collaboration, A. Evans (University of Virginia, Charlottesville/NRAO/Stony Brook University), K. Noll (STScI), and J. Westphal (Caltech); **357b** NRAO/AUI/NSF; **357t** NASA/UMD/A. Wilson et al; **358b** NASA / ESA / W. Keel (University of Alabama), and the Galaxy Zoo team; **358t** NASA and The Hubble Heritage Team (STScI/AURA); **359b** NASA, ESA, and The Hubble Heritage Team (STScI/AURA); acknowledgment: B. Holwerda (Space Telescope Science Institute) and J. Dalcanton (University of Washington); **359t** X-ray (NASA/CXC/UMD/Hodges-Kluck et al): Radio (NSF/NRAO/VLA/UMD/Hodges-Kluck et al); Optical (SDSS); **362–363** NASA, ESA, and the Hubble SM4 ERO Team & ST-ECF; **365** NASA, ESA, and D. Evans (Harvard-Smithsonian Center for Astrophysics); **366–367** 2MASS/T. H. Jarrett, J. Carpenter, & R. Hurt; **368b** HST/Frédéric Courbin & Pierre Magain; **368t** NASA, ESA, ESO, Frédéric Courbin (Ecole Polytechnique Federale de Lausanne, Switzerland) & Pierre Magain (Universite de Liege Belgium); **368–369** NASA,N. Benitez (JHU), T. Broadhurst (Racah Institute of Physics/The Hebrew University), H. Ford (JHU), M. Clampin (STScI), G. Hartig (STScI), G. Illingworth (UCO/Lick Observatory), the ACS Science Team and ESA; **369b** NASA/JPL-Caltech/ESA/Institute of Astrophysics of Andalusia, University of Basque Country/JHU; **369t** NASA, N. Benitez (JHU), T. Broadhurst (Racah Institute of Physics/The Hebrew University), H. Ford (JHU), M. Clampin (STScI), G. Hartig (STScI), G. Illingworth (UCO/Lick Observatory), the; ACS Science Team and ESA; **370t** NASA, ESA, C. Heymans (University of British Columbia, Vancouver), M. Gray (University of Nottingham, U.K.), M. Barden (Innsbruck), and the STAGES collaboration; **370b** X-ray: NASA/CXC/CfA/ M.Markevitch et al.; Lensing Map: NASA/STScI; ESO WFI; Magellan/U.Arizona/D.Clowe et al. Optical: NASA/STScI; Magellan/U.Arizona/D.Clowe et al.; **371** NASA, ESA, CXC, STScI, and B. McNamara University of Waterloo), NRAO, and L. Birzan and team (Ohio University); **372t** NASA/JPL-Caltech/Yale Univ.; **372br** NASA, ESA, ESO, Frédéric Courbin (Ecole Polytechnique Federale de Lausanne, Switzerland) & Pierre Magain (Universite de Liege Belgium); **372l** NASA, ESA, A, Bolton (Harvard-Smithsonian CfA), and the SLACS Team; **372tr** NASA/JPL-Caltech/Yale Univ.; **373br** NASA, ESA, and R. Gavazzi and T. Treu (University of California, Santa Barbara); **373l** NASA, ESA, M. J. Jee and H. Ford (Johns Hopkins University); **373tr** NASA, ESA, and R. Gavazzi and T. Treu (University of California, Santa Barbara); **374bl** NASA/JPL-Caltech/CXO/WIYN/Harvard-Smithsonian CfA; **374br** NASA/CXC/Penn State/G. Chartas et al; **374t** NASA, ESA, and the Hubble SM4 ERO Team & ST-ECF; **375** Andrew Fruchter (STScI) and NASA; **377** Springel et al (2005); **378b** X-ray: NASA, CXC, INAF, S. Andreon et al.; Optical: DSS, ESO/VLT; **378t** NASA/JPL-Caltech/Subaru/C. Papovich (Texas A&M Univ.); **379bl** Digitized Sky Survey (DSS), STScI/AURA, Palomar/Caltech, and UKSTU/AAO ; **379ml** NASA, ESA, R. Windhorst, S. Cohen, M. Mechtley, and M. Rutkowski (Arizona State University, Tempe), R. O'Connell (University of Virginia), P. McCarthy (Carnegie Observatories), N. Hathi (University of California, Riverside), R. Ryan (University of California, Davis), H. Yan (Ohio State University), and A. Koekemoer (Space Telescope Science Institute); **379t** 2MASS, T. H. Jarrett, J. Carpenter, & R. Hurt; **380b** NASA/JPL-Caltech; **380t** NASA/JPL-Caltech; **380–381** NASA/JPL-Caltech/Subaru/C. Papovich (Texas A&M Univ.); **381** NASA, ESA/JPL-Caltech/B. Mobasher (STScI/ESA); **382b** CXC/M.Weiss; Spectrum: NASA/CXC/N.Butler et al.; **382t** NASA, ESA, A. Straughn, S. Cohen and R. Windhorst (Arizona State University), and the HUDF team (STScI); **383** NASA, ESA, R. Bouwens amd G. Illingworth (University of California, Santa Cruz); **384–385** NASA, ESA, S. Beckwith (STScI) and the HUDF Team; **386b** Planck image: ESA/LFI & HFI Consortia; XMM-Newton image: ESA; **386t** NASA/CXC/; Penn State/G.Chartas et al; **387b** Springel et al (2005); **387tl** Springel et al (2005); **387tr** MacFarland, Colberg, White (Munchen), Jenkins, Pearce, Frenk (Durham), Evrard (Michigan), Couchman (London, CA) Thomas (Sussex), Efstathiou (Cambridge), Peacock (Edinburgh)/National Science Foundation/NASA; **388–389** NASA/WMAP Science Team; **390b** NASA, ESA, and R. Massey (California Institute of Technology); **390t** S. Columbi (IAP); **391** NASA/JPL-Caltech/E. Daddi (CEA France); **392l** NASA/JPL-Caltech/A. Kashlinsky (Goddard Space Flight Center); **392–393** Jodrell Bank Centre for Astrophysics, Univ. of Manchester; **393** NASA-Kepler team

Quercus Publishing has made every effort to trace copyright holders of the pictures used in this book. Anyone having claims to ownership not identified is invited to contact Quercus Publishing.

Quercus Publishing Plc
21 Bloomsbury Square
London
WC1A 2NS

First published in 2011

Copyright © 2011 by Quercus Publishing Plc

A catalogue record of this book is available from the British Library

UK and associated territories 978-0-85738-341-9

Canada 978-1-84866-136-3

Designed and edited by BCS Publishing Limited, Oxford.

Printed and bound in China

10 9 8 7 6 5 4 3 2 1